Make: 3D Printing

Compiled by Anna Kaziunas France

MAKER**MEDIA**™
SEBASTOPOL, CA

Make: 3D Printing

Compiled by Anna Kaziunas France

Maker Media books may be purchased for educational, business, or sales promotional use. Online editions are also available for most titles (*http://my.safaribooksonline.com*). For more information, contact O'Reilly Media's corporate/institutional sales department: 800-998-9938 or *corporate@oreilly.com*.

Editor: Brian Jepson

Production Editor: Christopher Hearse

Proofreader: Rachel Head

Indexer: Judith McConville

Cover Designer: Jason Babler

Interior Designers: Nellie McKesson and Brian Jepson

Illustrator: Rebecca Demarest

November 2013:   First Edition

Revision History for the First Edition:

2013-11-13:   First release

See *http://oreilly.com/catalog/errata.csp?isbn=9781457182938* for release details.

ISBN: 978-1-457-18293-8

[TI]

Table of Contents

Part II.   Software

Part III.   3D Scanning

6. Creating and Repairing 3D Scans 61

7. Print Your Head in 3D! 87

Part IV.   Materials

Part V.   Services

Part VI. Finishing Techniques

Part VII.   Applications

Part VIII.   Other Ways to Make 3D Objects

Preface

Additive Personal Fabrication

Written by **Anna Kaziunas France**

Welcome to *Make: 3D Printing*. I've compiled the best projects, tutorials, and stories about 3D printing from MAKE's print and online publications, refreshed them for the latest developments, and added a few new pieces you haven't seen elsewhere.

Personal Fabrication

This book contains stories and tutorials from and about makers who have embraced *additive personal fabrication*, a term that encompasses 3D printing activities at home, work, or school. Some of these makers are exploring not only DIY 3D printing, but creating scalable small businesses through the use of fabrication services for small batch custom manufacturing. Others create just what they need, just in time, from their desktop factory.

> As it turns out, the "killer app" in digital fabrication, as in computing, is personalization, producing products for a market of one person.
>
> — Neil Gershenfeld
> *Foreign Affairs November/December 2012*

Although additive manufacturing has been the subject of some irrational media exuberance, it's only part of the digital fabrication equation. As an instructor at the Fab Academy, a globally distributed rapid prototyping course where we turn codes into things, I taught a wide variety of digital fabrication techniques, often using subtractive methods and machines to achieve desired results.

This in no way diminishes the opportunities for creativity and expression that are made possible by 3D printing. Through 3D scanning and printing I have been able to quickly model and fabricate a variety of completely bespoke (and often last minute) items; from costumes to artwork to functional molds.

Part of the power of personal fabrication is directly tied to having easily accessible personal machines. Machine access combined with creativity and time pressure can lead to interesting mashups and materials choices.

When working on the scanning chapter of *Getting Started with MakerBot* (see Chapter 6), I created a scan of a prop skull. I later decided to use the scan to create the traditional necklace of skulls or "garland of Kali" as costume for Halloween. Because I had two desktop printers at my home, I was able to prototype my sizable Kali necklace in a single week by running them continuously. I also used the skull scan to create batches of molded skull chocolates (see the Chapter 21 tutorial).

A month later, the Kali necklace was joined by a belt and I incorporated it into a four armed sculpture I titled "Self Portrait as Kali" (see Figure P-1). I created the body of the sculpture from several self scans that I combined digitally to create two sets of arms on a single torso. Both sets of arms are broken off at the forearms, as if the original scan was of an ancient statue that has been damaged over time. The final 3D model was sliced into 125 individual routed slices of 1/2" medium density fibreboard (MDF) that I fabricated on a large CNC (computer numerically controlled) router. I then assembled and painted the slices by hand. The belt and necklace, I also painted by hand and strung them over the assembled body. The finished sculpture has been displayed at several exhibitions, including the "Bits to Its" 3D printed sculpture show and the RISD Museum.

In many ways, 3D printing is currently the most personally accessible of all types of digital manufacturing. Prices have come down enough for many to afford to own their own printers, and desktop machine print quality has dramatically improved. Online printing services are readily available for those who cannot afford printers or whose materials needs go beyond printed plastic.

The world of personal fabrication is rapidly evolving—and you are now part of it!

Figure P-1. *Self Portrait as Kali*

Who This Book Is For

If you are interested in creating your own one of a kind or small batch customized creations using 3D printing processes, this book is for you.

If you're absolutely new to 3D printing and don't know a thing about it, you'll want to read this book cover to cover.

If you've already got a 3D printer, but are ready to go beyond just printing designs you downloaded from the Internet, you'd want to start by creating your own printable objects by learning 3D design with Tinkercad and creating 3D scans. Also check out the finishing section for tips on how to refine the appearance of your 3D prints.

Are you more comfortable with software than with hardware? If you are a designer who wants to prototype in a range of materials other than extruded plastic or a desktop 3D printer is out of your budget, check out the services and materials sections for a rundown of the range of companies that will print your designs in a rapidly growing number of exotic materials.

Contents of This Book

Part I is an overview of the 3D printing hardware; the printers themselves, the basics of how they work, and what to expect in your experiences with them.

In Part II, you'll learn about the software toolchain required to take 3D printed designs to final printed object and how to design your own objects.

Part III takes 3D model creation a step further with several tutorials on how to capture physical objects around you with 3D scanning. You'll also learn how to clean the scanned models for 3D printing.

Part IV discusses a the ever-expanding range of plastic filament available for desktop 3D printing. You'll also learn about industrial printing materials (and methods) for ceramics to metals available from 3D printing services.

Part V deals with how and why to use 3D printing services instead of desktop manufacturing and details about the different service provider options available.

Several methods for finishing your for 3D printed objects are covered in Part VI, including changing the color or your prints with fabric dye, repairing broken prints and techniques to "weather" your prints to look like battered metal.

Part VII explores the plethora of possible personally fabricated creations that are possible through additive manufacturing; from humanoid robots to scanned artwork to fully customized prosthetics.

Part VIII describes other ways to create 3D objects, from milling to creating food-safe molds for chocolate casting.

Anna Kaziunas France is the Digital Fabrication Editor of Maker Media. She's also the Dean of Students for the Global Fab Academy program and the co-author of Getting Started with MakerBot. *Formerly, she taught the "How to Make Almost Anything" rapid prototyping course in digital fabrication at the Providence Fab Academy. Learn more about her at* her website (http://kaziunas.com) *and check out her things at* her Thingverse page (http://thingiverse.com/akaziuna).

Conventions Used in This Book

The following typographical conventions are used in this book:

Italic

> Indicates new terms, URLs, email addresses, filenames, and file extensions.

`Constant width`

> Used for program listings, as well as within paragraphs to refer to program elements such as variable or function names, databases, data types, environment variables, statements, and keywords.

`Constant width bold`

> Shows commands or other text that should be typed literally by the user.

`Constant width italic`

> Shows text that should be replaced with user-supplied values or by values determined by context.

> *This section signifies a tip, suggestion, or general note.*

 This icon indicates a warning or caution.

Using Examples

This book is here to help you get your job done. In general, if this book includes code examples, you may use the code in this book in your programs and documentation. You do not need to contact us for permission unless you're reproducing a significant portion of the code. For example, writing a program that uses several chunks of code from this book does not require permission. Selling or distributing a CD-ROM of examples from MAKE books does require permission. Answering a question by citing this book and quoting examples does not require permission. Incorporating a significant amount of examples from this book into your product's documentation does require permission.

We appreciate, but do not require, attribution. An attribution usually includes the title, author, publisher, and ISBN. For example: "*Make: 3D Printing* (MAKE). Copyright 2014 Maker Media, 978-1-457-18293-8."

If you feel your use of code examples falls outside fair use or the permission given above, feel free to contact us at *bookpermis sions@makermedia.com*.

Safari® Books Online

 *Safari Books Online* is an on-demand digital library that delivers expert content in both book and video form from the world's leading authors in technology and business.

Technology professionals, software developers, web designers, and business and creative professionals use Safari Books Online as their primary resource for research, problem solving, learning, and certification training.

Safari Books Online offers a range of product mixes and pricing programs for organizations, government agencies, and individuals. Subscribers have access to thousands of books, training videos, and prepublication manuscripts in one fully searchable database from publishers like MAKE, O'Reilly Media, Prentice Hall Professional, Addison-Wesley Professional, Microsoft Press, Sams,

Que, Peachpit Press, Focal Press, Cisco Press, John Wiley & Sons, Syngress, Morgan Kaufmann, IBM Redbooks, Packt, Adobe Press, FT Press, Apress, Manning, New Riders, McGraw-Hill, Jones & Bartlett, Course Technology, and dozens more. For more information about Safari Books Online, please visit us online.

How to Contact Us

Please address comments and questions concerning this book to the publisher:

> Maker Media, Inc.
> 1005 Gravenstein Highway North
> Sebastopol, CA 95472
> 800-998-9938 (in the United States or Canada)
> 707-829-0515 (international or local)
> 707-829-0104 (fax)

We have a web page for this book, where we list errata, examples, and any additional information. You can access this page at *http://oreil.ly/make-3d-printing*.

To comment or ask technical questions about this book, send email to *bookquestions@oreilly.com*.

Maker Media is devoted entirely to the growing community of resourceful people who believe that if you can imagine it, you can make it. Maker Media encourages the Do-It-Yourself mentality by providing creative inspiration and instruction.

For more information about our publications, events, and products, see our website at *http://makermedia.com*.

Find us on Facebook:

https://www.facebook.com/makemagazine

Follow us on Twitter:

https://twitter.com/make

Watch us on YouTube:

http://www.youtube.com/makemagazine

PART I

Hardware

Getting Started with a 3D Printer

1

An introduction to 3D printer hardware and software.

WRITTEN BY **BILL BUMGARNER**

Early in 2012, I picked up an Ultimaker, put it together, and joined the growing ranks of 3D printing households. It has been an adventure both filled with reward and rife with frustration. The goal of this article is to share what I've learned while studying the DIY portion of the 3D printing realm. The focus is on budgets less than $2,500, with a goal of producing parts out of various kinds of plastic.

Plastic parts are wonderful for prototyping. You can print that engine part in plastic, make sure it fits perfectly, and then send the 3D model off to a company like Shapeways to have your prototype turned into a production piece in the metal of your choice.

Most of the printers discussed here are hackable. Their designs are amenable to being modified and tuned to fit your needs. The software used to drive these printers is almost all open source, though there are commercial slicers and modelers commonly used in the 3D printing community.

Choosing a Printer

The 3D printers discussed here are of the *additive manufacturing* variety. They create parts by adding material together and are the new hotness in the field of manufacturing. So hot, in fact, that the Obama adminis-

tration created the National Additive Manufacturing Innovation Institute (NAMII) to foster innovation in this field.

There are three approaches to additive manufacturing in common use: *photopolymerization* (using light to cure a liquid material into solids of the desired shape), *granular materials binding* (using lasers, hot air, or other energy sources to fuse layers of powder into the desired shape), and the focus of this article, *molten polymer deposition* or MPD (extruding molten material in layers to build up the desired shape).

In short, MPD, aka fused deposition modeling (FDM) or fused filament fabrication (FFF), describes pushing a filament of solid plastic (or other materials like metal or chocolate) into a hot-end that then extrudes a thin stream of molten material in layers to build up the desired piece.

Of these technologies, MPD is the most common and most accessible (though EMSL's CandyFab definitely bears mention because any printer that smells like crème brûlée deserves a shout-out).

Focusing on MPD, there are a handful of different styles of printers. The differentiation is largely focused on exactly how the printer moves the extruder to a particular point to extrude the plastic.

Buying Options: Turnkey, Kit, or DIY

With this information in hand, it's time to choose a printer! There are many turnkey solutions that go well beyond our targeted price range. And there are some that are quite affordable. The Up 3D printer (the same as the "Afinia H-Series" on page 18) is an example of a "ready-to-print" device that re-quires relatively little maintenance. The Mak-erBot Replicator is similarly focused, but is both more versatile and may require a bit more maintenance.

A word of warning: when buying a turnkey printer, be wary of "razor vs. blades" business models. 3D printers exist that are seemingly cheap, but which require proprietary fila-ment cartridges, where the consumable fil-ament costs two to three times the going market rates.

Plastic Prototyping

Bathsheba Grossman's beautiful (and incredibly pop-ular) Klein Bottle Opener is a perfect example of the prototype-in-plastic, print-in-metal process that al-lows for lots of cheap plastic drafts before commit-ting to the relatively expensive process of procuring a finished metal part.

On the left is a 3D print of Grossman's bottle opener. On the right is the same model printed in stainless steel and brass from Shapeways. You could tune and print the model all day long for very little money (less than $1 in plastic per print) and then commit to the metal version once fully satisfied with geometry, fit, functionality, etc.

2. Removing the support

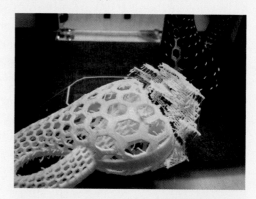

1. A draft print of the Klein Bottle Opener still attached to the print bed with the support visible at the bottom

3. The finished draft beside the final piece

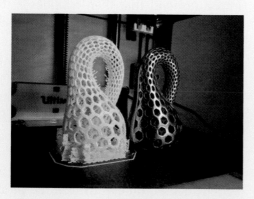

Moving to kits, both the Printrbot and Ultimaker are examples of printers that arrive as kits to be assembled by the end user. Both feature great instructions, and assembly is focused on mechanical construction, since both include all electronics pre-soldered and ready to go.

If you're interested in the full DIY experience, both the RepRap and the Rostock printers are entirely DIY. Both printers are composed of a combination of 3D-printed parts and various metal or wood bits, along with some electronics. You can order preprinted parts kits from a variety of sources (eBay included). Electronics and/or mechanical packs are available, too. Generally, the more cutting-edge the printer version, the more parts you'll need to source.

In "3D Printer Anatomy" on page 7, you'll find an overview of 3D printer anatomy, breaking down positioning systems, parts, and filament.

Software

The workflow for turning an idea into a 3D print can be summed up as *model* (or *capture*), *fixup*, *slice*, and *print*. At each step, there are multiple software solutions to choose from.

Generating STL files

STL files are the lingua franca of the 3D printing world. If an application can export a 3D model as an STL file, then that STL file can be sliced and printed. STL files can be generated using a *CAD program*. SketchUp (*http://sketchup.com*) is quite popular, as are a number of open source 3D modelers.

Regardless of which modeler you choose, expect the learning curve to be steep. The key challenge is to translate what you want

to make into a set of primitive shapes and elements in the modeler, such that the result is possible to print. For example, any overhang shallower than about 45° cannot be printed without support material (which consumes material that will be tossed, increases print time, and requires significant cleanup) because the plastic will sag (the actual angle varies from printer to printer and depends on how you configure the slicer).

Alternatively, models can be generated using a *parametric CAD program*, of which OpenSCAD (*http://www.openscad.org/*) is quite popular. Instead of drawing what you desire, you write code in a simple language that tells the CAD system what you want.

Many of the mathematics-targeted packages like Mathematica can also produce STL files.

Models can also be generated from photographs or videos. Autodesk's 123D Catch (*http://123dapp.com/catch*) can process a series of photos and turn them into a 3D model (see Chapter 6). With an iPhone or iPad, you can take up to 40 pictures of a model using 123D and then upload them to a server that processes the images into a 3D model. Note that Autodesk also released 123D Make (*http://123dapp.com/make*), which slices 3D models into plans for constructing the model out of cardboard or paper.

Often the model for what you want has already been created by someone who shared it online. The most popular repository for pre-made models (and full-on projects) is Thingiverse (*http://thingiverse.com*), where you can find thousands of printable solutions for everyday problems: phone cradles, earbud holders, silverware dividers, quadrocopters (yes, printable quadros), camera parts, game pieces, printable 3D printers, etc. You name it, there are likely at least a few good starting points on Thingiverse.

Slice

The *slicer* is what turns an STL file into a series of commands—typically G-code—that tell the printer where to move the print head and when to extrude plastic. While the printer's driver software is dictated by the printer, it's actually a critical piece, as the printer software will often model the motion dictated by the stream of commands and will vary the acceleration of the motors to eliminate print artifacts.

Slicing is a critical phase of the print. It's a careful balance between quality, speed, and amount of material used. In many cases, the choice of appropriate slicing parameters is the difference between a successful print and a pile of spaghetti.

Fixup

During *fixup*, the STL to be printed is often checked for errors (modelers sometimes dump 3D descriptions that look OK, but can't translate directly to a 3D print because the structure is invalid). The model may be rotated or scaled, or it may be duplicated to print multiple copies simultaneously. Often, slicing and fixup are contained in the same program (sometimes with the printer driver, too). Cura (*http://wiki.ultimaker.com/cura*) and Slic3r (*http://slic3r.org*) combine slicing with limited fixup, offering the ability to rotate, scale, and print multiple different STL files in a single print run. Netfabb (*http://netfabb.com*), a commercial package, has extensive mesh debugging and fixup capabilities, along with basic layout tools and a powerful slicer (see "Cleaning and Repairing Scans for 3D Printing" on page 70).

Five 3D-Printed Replacement Parts for My Ultimaker

Five 3D-printed replacement parts for my Ultimaker: a new drive gear, the white knob holding the gear on, the Bowden tube clamp (the white piece at the top of the extruder assembly), the orange piece on the left that holds the extruder in place, and a filament spool holder (hidden).

Your First Print Job

Now that you have a background on all things 3D printing and have bought or built your first 3D printer, how do you go about creating useful things (or creating useless, but very cool, things)?

The first step is to know your tools. If applicable, download and print all of the upgrades and replacement parts for your printer. For the Ultimaker, I've printed a number of parts, some of which can be seen in the image shown in the sidebar "Five 3D-Printed Replacement Parts for My Ultimaker."

Then download and print some useful things from Thingiverse. Start with small items—but not too small—like an earbud holder, a bottle opener, or a simple character. Choose something where lots of folks have uploaded photos of their own versions (under the "Who's Made It?" section). By starting with a known, working model with lots of examples, you can compare your product to others and will have a much better basis for fixing any problems and tuning your workflow (likely the slicer, in particular) to maximize print quality from your printer.

Each type of 3D printer has a particular sound of a successful print. Learning it can help you identify problems before they become serious. I can be in a different room and still tell when a print on my Ultimaker is about to fail just because it starts to sound out of whack.

Have a look at how various models are constructed. The key challenge in creating a 3D model is how the desired real-world object can be broken down into a series of commands—"draw a line and a curve," "extrude this surface," "fillet that corner," "cut a hole here"—that you can use to create the object. If you can find some models on Thingiverse that are in a file format you can edit, try your hand at editing them to see if you can add a feature you desire.

Once you're ready to start from scratch, the first tool you'll need is a set of calipers. Digital slide calipers with 0.01 mm accuracy can be had for less than $20. Try creating hangers or hooks that require a precise measurement to fit snugly on an already existing object (like over a door), giving both a feel for using the calipers and helping you dial in how your printer maintains dimensions throughout the printing process.

One unique challenge is learning to think in a 3D-printer compatible way. As mentioned, objects with structures that tilt at more than 45° can't be printed without support material because the object will sag. Note that

bridges—linear spans of plastic over a gap—work surprisingly well, but print them so the first layer of the bridge is on the inside of an object and won't be seen.

While you can turn on "support material" in most slicers, doing so often causes a lot of extra material to be used, which is both wasteful and requires a ton of post-print cleanup. Note that with a dual extruder printer you can print PLA or ABS on one extruder while printing with water-soluble PVA on the other extruder, making cleanup a simple matter of soaking the piece in a bucket overnight.

There is also a balance between creating walls that are too thinly and too thick. Thin walls can lend elegance to an object, but print too thin and that single or double layer of linear plastic filament will fail quickly. Naturally, this all becomes more intuitive with practice and experience.

3D Printer Anatomy

There are a lot of parts, moving and otherwise, that you'll find inside a 3D printer.

3D Positioning System

Within the build volume, the printer must be able to position the hot-end at any point to be able to precisely extrude material within the layer being printed. However, motion doesn't have to be limited to the hot-end.

3D printers may move the hot-end or the print bed in a number of different ways. Here are three primary designs in use today:

Gantry
> This style of printer moves the extruder in the x- and y-axes, while the bed moves only in the z-axis. Both the Ultimaker (Figure 1-1) and the MakerBot Replicator (Figure 1-2) use a gantry to move the

extruder. However, the similarity ends there. The Replicator integrates the extruder with the hot-end. The Ultimaker uses a Bowden cable to separate the two. The Replicator has a simpler design, whereas the Ultimaker greatly decreases the weight of the print head, allowing for greater print speeds (at the cost of additional maintenance issues).

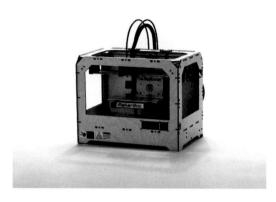

Figure 1-2. *The Replicator*

Moving bed

Instead of moving the print head in xy space, one of the axes is changed by moving the print bed itself. Usually, a moving-bed printer will move the print head in the z-axis. This is a mechanically simpler design in that the x- and y-axes are managed independently using entirely linear motion. It has the disadvantage of requiring the printer to move a significantly heavier print bed, which could knock loose the printed part. It's slower, but simpler. Printrbot's models (Figure 1-3) are examples of moving-bed printers that trade print speed for low cost and ease of maintenance.

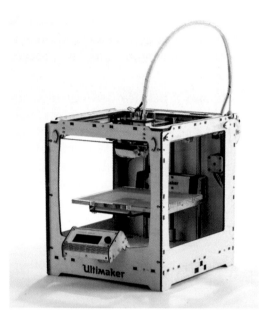

Figure 1-1. *The Ultimaker*

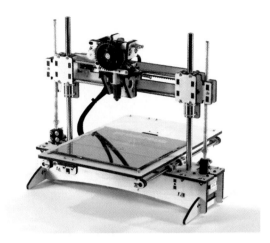

Figure 1-3. *A Printrbot*

Deltabot

Industrial pick-and-place robots typically use this design. A deltabot has three control rods connected to the toolhead, and these rods can be moved to control the position of the head. Recently, Johann Rocholl adapted this technology to 3D printing and created the Rostock (Figure 1-4). Hard to explain, it looks like an alien probe is printing your favorite model.

The Rostock printers (so far) use a Bowden setup to separate the hot-end from the extruder, allowing for very quick and precise head positioning with relative mechanical simplicity. The downside is increased complexity in the driver; the hot-end positioning is not a linear set of steps because of the non-linearity of the motion between the vertical axis and the planar positioning of the hot-end.

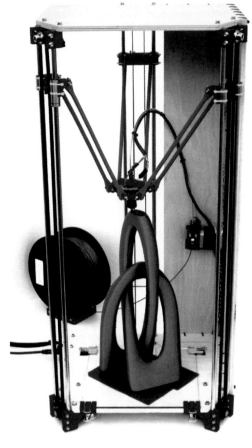

Figure 1-4. *The Rostock*

3D Printer Parts

Print bed

This is the bed upon which the printed part rests during production. Bed temperatures can be ambient or heated. A non-heated bed is often covered in painter's tape, as seen in the Ultimaker (Figure 1-1), to which the printed material adheres. Heated beds, as seen in the Printrbot (Figure 1-3), keep the part warm during the print and prevent warping. Depending on the material, a heated bed will maintain a temperature from 40°C to 110°C throughout the print.

Let it be noted that the insides of a 3D printer are not a finger-friendly zone (and I have the blisters to prove it)!

Extruder

The extruder is not actually the part that squirts out plastic. The extruder is the part that feeds the plastic filament into the hot-end. Extruders may be integrated into the hot-end or they may be remote, typically pushing the filament through a stiff PTFE (Teflon) tube (this is the Bowden cable) into the hot-end (Figure 1-5).

With a dual extruder (Figure 1-6), you can print two different materials or colors simultaneously. This versatility comes at a cost of complexity (and price), as it requires an extra extruder, hot-end, and all the bits in between. Some printers, like the Ultimaker, can be upgraded from single to multiple extruders. Others cannot.

Figure 1-6. *A Replicator sporting a dual extruder*

Hot-end

The hot-end is comprised of a heater, a temperature sensor, and an extrusion end through which the plastic filament is pushed to deposit molten material (Figure 1-7). Hot-ends are often assembled within an aluminum block or are configured in a barrel-type shape.

Note that the interface between the hot-end and the extruder—be it directly integrated or with a Bowden cable between the two—can be exceptionally problematic on some printers or on a printer that is not adjusted properly.

Figure 1-5. *A Bowden extruder*

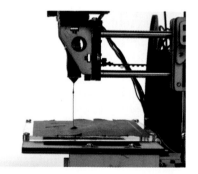

Figure 1-7. *The Printrbot hot-end / extruder*

The hole in the nozzle may range in size, typically between 0.2 mm and 0.8 mm.

The smaller the nozzle, the more detailed the print, but the longer it takes.

Plastic filament

The filament is the consumable of the printer. Like an inkjet squirts ink, a 3D printer squirts melted filament.

Choosing a Filament

There are a handful of plastics that are compatible with MPD-based printers. Each has its pros and cons. I choose to print exclusively with PLA because of its low toxicity and general environmental friendliness.

All printers are not compatible with all materials. Some materials may really stress a printer, as the temperature range for extrusion may vary from 160°C to 305°C, depending on the material. A printer designed for PLA/ABS at a max of 250°C may fail entirely at 300°C.

With that in mind, the three most common materials are PLA, ABS, and PVA. (Read more about materials choices in Chapters 8 and 9.)

ABS

ABS (acrylonitrile butadiene styrene), the cheapest of the three, is typically extruded at between 215–250°C, and does best with a heated bed to prevent warping. ABS creates mild, generally tolerable fumes that may be dangerous to sensitive people or certain pets (personally: fume hood, please!). ABS can be quite versatile. It can be sanded, and by mixing ABS with acetone, it can be easily glued together or smoothed to a glass-like finish.

PLA

PLA (polylactic acid or polylactide) is a biodegradable plastic typically made from corn or potatoes. PLA filament is extruded at a lower temperature of 160–220°C and does not require a heated bed (painter's tape is just fine). When heated, PLA smells a bit like sweet, toasted corn. PLA tends to be stiffer than ABS. While PLA does not require a heated bed, it can warp a bit during cooling, something that a heated bed can greatly improve. Note that there is a "flexible PLA" variant that, while trickier to use, will result in objects that are squishy.

PVA

PVA (polyvinyl alcohol) is a specialty plastic used on multiple-extruder printers to print support material. PVA is typically extruded at 190°C, is water soluble, and can be used to print support material in complex 3D prints with lots of overhangs. PVA absorbs water like a sponge, which causes problems in high-humidity environments.

Alternative Materials

While MPD printers are generally tuned to filament printing, it's not hard to adapt them to support other materials. A common mod is to add a syringe-style extruder that handles materials such as chocolate, frosting, and various kinds of clays.

Next Steps: What to Make

Once you start to get the hang of 3D printing, a world of possibilities opens up. I've given many of these things away as gifts. The material costs are very cheap and the look of surprise when you say, "Yeah, I printed that. You want it? Different color? Nah, no problem, I can print as many as you want easily," is priceless! With my Ultimaker, I have printed the following handy bits, and many, many more (see Figures 1-8 through 1-15).

Figure 1-8. *Replacement screw cap for a large bottle of Jack Daniels (http://thingiverse.com/thing:18194)*

Figure 1-9. *Ultimaker tool holder (http://thingi verse.com/thing:18098)*

Figure 1-10. *Raspberry Pi case (http://thingi verse.com/thing:25363)*

Figure 1-11. *Nautilus-shaped earrings (http://thingi verse.com/thing:13450)*

Figure 1-12. *Aeroponic grow pods for a winter herb/ salad garden (http://thingiverse.com/thing:32613)*

Figure 1-13. *Fan shroud with spacer(http://thingi verse.com/thing:16530)*

Figure 1-14. *Case for a Teensy-based IR Blaster (http://thingiverse.com/thing:19315)*

Bill Bumgarner (http://friday.com/bbum) plays with high voltage, cooks with fire and water, incubates microbes, hacks code, corrals bugs with his son, and tries to make stuff do things that were never intended.

Figure 1-15. *Drawer divider for my son's screw sorting efforts (http://thingiverse.com/thing:32614)*

3D Printer Guide

2

Hands-on experiences with 11 3D printers.

In September 2012, *MAKE* invited Matt Griffin to put together a team of reviewers from the 3D printing community to advise *MAKE*'s readers about the 3D printing state of the art.

The team investigated 15 of the most promising fused-filament fabrication (FFF) printers on the market—devices that melt and extrude plastic filament to form solid objects, layer by layer—the most popular method of 3D printing. We reached back to 2010 and into the future, previewing still-under-wraps machines weeks before their launch at World Maker Faire New York.

A lot has changed since those tests: there are new printers on the market, some printers are no longer for sale, and many have improved. This chapter distills the original reviews into stories of hands-on experiences with the 11 3D printers that, at the time of this publication, are still for sale. If you want the most up-to-date reviews, check out the latest edition of our *Make: Ultimate Guide to 3D Printing* (*http://makezine.com/volume/make-ultimate-guide-to-3d-printing/*); if you want to learn what it's like to explore a printer from the moment it comes out of the box, read on!

Figure 2-1. *During the testing, the MAKE offices became a vortex of 3D printers and the geeks who love them*

The Challenge Prints

Here are the objects we used in our testing.

Snake

by Zomboe (*http://thingiverse.com/thing: 4743*)

This reinterpretation of a classic wooden toy features flexible ribs that are a great test for both horizontal accuracy on the plate (are the ribs evenly spaced and complete out to their tips?) and vertical registration (does each layer match perfectly with the one beneath?).

Owl Statue

by Tom Cushwa (*http://thingiverse.com/ thing:18218*)

Designer Cushwa borrowed stone-cutting techniques to render feathers and character features for this popular owl figurine and modeled them to look great on a well-tuned printer. Printers that choke on these details may not be suitable for character and sculptural work.

Nautilus Gears

by Misha T. (*http://thingiverse.com/thing: 27551*)

This model gives character to the classic snap-together gears that are a popular test object for 3D printers. It's quick to print, and you can tell instantly how well the machine reproduced the parts from how accurately the teeth mesh and the snaps snap, and whether the gears can rotate through more than one revolution without binding.

Dimension Torture Test

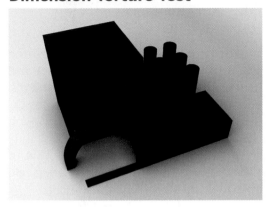

by Cliff L. Biffle (*http://makezine.com/3dprint ing*)

To create a real "torture test"—a model guaranteed to challenge all FFF printers—engineer Cliff L. Biffle built a Frankenstein's monster containing all the geometry we needed to see in one small package. Thin and fat features, slopes and overhangs, bolt holes with precise dimensions, arcs, and separate towers all conspire to push a machine to its limits.

Afinia H-Series

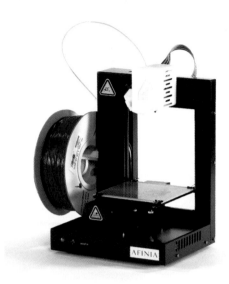

- *http://afinia.com*
- Written by Keith Ozar
- Tested by Keith Ozar and Eric Weinhoffer

The compact H-Series is a version of PP3DP's UP! Plus 3D printer, rebranded for the US market. It's got a single extruder and runs via USB off Mac or Windows machines, with no onboard controls except an initialize button and a flashing status indicator. But despite the no-frills hardware, it's a surprising little performer that's ideal for beginners.

The printer sits about 10"×10"×14" high on the desktop, weighs just 11lbs, and is sturdy for transport. It ships fully assembled and takes only a few minutes to set up. Loading the filament, leveling the platform, and calibrating the machine were easy, thanks to its straightforward documentation.

The Afinia 3D proprietary software package slices, generates support, duplicates, and can place multiple models for printing—all automatically. Once your design is to your liking, it's as easy as pressing Print.

Print quality was quite good—we were surprised by how great the first print looked (the snake). And subsequent prints like the owl confirmed it: sharp corners, clean overhangs, and true vertical and horizontal surfaces. Dimensional accuracy was off by about 1% on our torture-test object, with holes slightly undersized, though this might be due to shrinkage of ABS plastic. The Afinia was reliable, too; we didn't have to babysit it.

Unfortunately, there is no SD card or USB flash drive on the Afinia. You can print from your computer via USB and disconnect once the job has started printing, though.

The status light indicators can be confusing if you don't have the manual in front of you, and are a constant reminder of the lack of an onboard control panel. The machine also makes a loud beeping sound as it begins printing, which reminded us of a truck backing up, though it prints quietly.

Though it printed slower than many of the printers we tested, the H-Series stood out as one of our favorites. Straight out of the box, using the default 0.2 mm resolution settings, we printed some of the best-looking challenge prints.

Bukobot 8

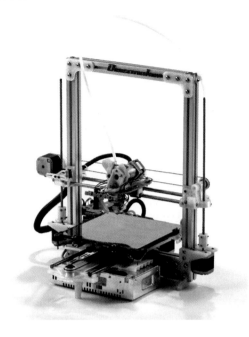

- *http://deezmaker.com*
- Written by Matt Griffin
- Tested by Matt Griffin and John Abella

We tested the Bukobot 8 Vanilla version, with a single extruder; the Duo version offers two extruders. Setup took only about a half hour, and the documentation included with the printer provided most of the information necessary to tune the machine, though we did have to seek help to figure out the baud rate.

A great touch: printed cable tensioner disks make it easy to perform small dialing-in adjustments. The version we tested had no SD card reader, but the current version includes it for untethered printing. The Budaschnozzle hot-end has an idler latch for easy filament swapping mid-print.

While we felt confident the gantry was square to the plate, we had adhesion troubles with both ABS and PLA until we covered the platform with blue painter's tape—a common solution out in the field for printers that extrude PLA. This improved adhesion significantly, and we were able to print a fine-looking snake using Printbl's new Diamond Age PLA, which Deezmaker included with the unit.

After shifting to the printed fan shroud attachment, which allows for active cooling of the top printed layer, we had our best successes printing PLA. The torture-test results were promising: adhesion issues knocked off the base of the unsupported arch, but the vertical and horizontal surfaces were all true.

Printing results generally were better than average, and the owl statue in particular, while not the best in our test, promises that this unit, when really dialed in by an operator, will compete aggressively for quality at its price point.

Cube

- *http://cubify.com/cube*
- Written by John Abella
- Tested by John Abella and Matt Griffin

The first thing you notice about the Cube is its style. No plywood, loose wires, or zip ties: it's more a consumer appliance than a hacker plaything. This level of design quality carries into the included documentation, which was among the most thorough.

The Cube brings a mix of unique features as well, including the ability to print via WiFi or USB, a well laid-out touchscreen control panel, and a completely new item that no other manufacturer had: "Magic Cube Glue."

The Cube can use your WiFi network, but we had better luck printing via its ad hoc network. You'll have to disconnect briefly from any other WiFi network to use it, though.

Like most ABS printers, the Cube uses a heated build platform to help prints stick and to prevent warping. Also, the included "magic

glue" worked great to keep prints stuck to the build surface. This mystery substance washes away with water to cleanly release prints from the removable platform.

The device is driven by a touchscreen interface that allows the user to configure nozzle height, WiFi settings, and other details, as well as showing the status of current print progress and other system indicators.

Out of the box, the Cube was a bit slow. By default it enables rafts and support structures, both of which add to printing time and post-processing time, but which helped in the most demanding print tests. Both options can be turned off, but doing so will generally degrade the print quality. While many 3D printing software packages allow for endless tweaking, the Cubify package has basic switches for toggling support and rafts, but not much control over other settings.

The Cubify software works with standard STL files but also supports proprietary *.creation* files. 3D Systems has a website where you can buy creations. Each new printer comes with 25 free files, and some additional designs are free. The Cube works with filament cartridges available only from 3D Systems. At $50 for about 1 pound of usable ABS, they are pricier than other suppliers.

Felix 1.0

- *http://felixprinters.com*
- Written by Eric Chu
- Tested by Eric Chu and Brian Melani

The Felix is based on RAMPS 1.4 electronics, has a rigid frame made of aluminum extrusion, and has a generous build volume. It took us longer than the estimated 5–10 hours to finish this build, and we've built kit printers before.

The Felix version of Repetier bundles Skein-forge and Slic3r under the Repetier frontend, so you can choose one or the other when slicing your STL file. Felix came with a handful of slicing profiles, as well.

Leveling the bed and setting the Z height were a bit of a challenge. You use a wrench to loosen a nut under each of the three leveling screws until the aluminum bed is leveled.

The Z limit switch is an optical endstop that's triggered when an opaque object blocks its IR beam. An LED on it should turn off when the endstop is triggered, but ours only got very dim, making it hard to know exactly where the Z home position was triggered. However, the X and Y belts are extremely easy to tension: just turn one screw on each axis to tighten them.

The Felix's extruder uses a spring-loaded tensioning mechanism that presses a ball bearing against the filament and extruder gear. We finally got the proper tension when we set it so that the ball bearing rests against the extruder gear before putting in the filament.

The Felix is a pretty speedy machine. Its print quality is great when the belts and Z lead screw are properly tensioned and aligned. Aligning the Z lead screw is the hardest part of getting a great print; if it's off, there will be ridges between the layers, an effect known as *Z wobble*.

The Felix was designed to print in PLA. While it has a heated bed, its open-case design doesn't allow the bed to get hot enough for printing ABS. The bed also needs to be slightly leveled every couple of prints, and it can warp due to the flexing of the aluminum.

Felix's best feature is that it's extremely quiet. DryLin polymer bushings glide along the linear rails so stealthily that even at high speeds, it's quiet.

MakerGear M2

- *http://makergear.com*
- Written by Paul Leonard
- Tested by Paul Leonard

This printer's frame is stainless steel, and the rest of the structural elements are stainless and anodized aluminum. The print volume is 8"×10"×8"—one of the biggest in its price range. The rest of the printer is built with premium parts, and its motion is smooth.

MakerGear's geared stepper motor extruder comes with their groove-mount hot-end. The heated build platform is a sandwich of cork insulation, polyamide heating element, and laser-engraved aluminum, topped off with a sheet of borosilicate glass.

UltiMachine's RAMBo board, a refinement of the popular RAMPS, controls the printer. It supports dual extruders, so you could upgrade to dual-extruders without upgrading the electronics.

The kit is very well organized and includes every tool you need except an adjustable wrench. Kit documentation is very nice—animated videos, subassembly drawings, explicit instructions for critical steps—but also has confusing gaps. That said, MakerGear tech support answered questions quickly on IRC.

Setup is simple: you can unpack, adjust the stops, load the software, and start printing the included samples in less than an hour. Find a sturdy table—the M2's metal construction makes it reassuringly heavy, and it shakes a bit when it runs at high speeds.

The M2 ships with an SD card reader and a card loaded with sample G-code files to verify that your printer is working. They also include a 1 kg spool of PLA. Although it ships well calibrated, there's a learning curve. Before you print, visit MakerGear's Google Groups to learn about the settings and options available, especially for different materials and build surfaces.

After I positioned the Z endstop and leveled the bed, the M2 printed great. I experimented with different layer heights, speeds, and temperatures, and I've yet to experience the M2's full capabilities. Not only can I adjust the software settings to get even better output, but because the machine accommodates improvements in extrusion technology, I'll be able to keep on upgrading!

Printrbot Jr. (v1)

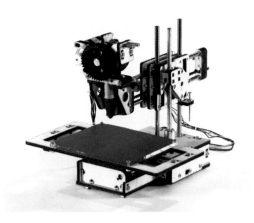

- *http://printrbot.com*
- Written by Lyra Levin and Matt Griffin
- Tested by Lyra Levin, Cliff L. Biffle, Emmanuel Mota, and Blake Maloof

One of the tiniest and least expensive 3D printers, the Jr. is a baby-scale version of the original Printrbot. (Some of the other printers could probably print a full-scale model of the Printrbot Jr. in one go.)

When fully open, the diminutive machine fits easily on a corner of your desk, and most of its footprint is the 4"×4"×4" printing volume. When folded for travel, a good third of the volume is packed in, with most of the fiddly bits protected by the bed. It easily slips into a backpack.

The unit comes with a standard PC ATX power supply, but there's another option for power: a rechargeable lithium polymer (LiPo) battery as an optional add-on to supply power when printing in the field. We ran some of our final test prints using a LiPo quadcopter battery (soldering required), and were amazed to realize we could just as easily be printing them on top of a mountain somewhere.

Simplicity is a design feature for this printer, and while it ships assembled, videos of Drumm quickly disassembling and reassembling the unit demonstrate how carefully the machine has been reduced to the fewest moving parts possible.

In his quest for simplicity, however, Drumm sacrificed some bells and whistles: the Jr. prints only PLA, doesn't have a heated platform or a fan, and has a simplified gantry system.

The first prints we rolled off the Jr. were officially "not bad." Following these, we took time to carefully level the platform by eye, running the nozzle horizontally across the plate to make sure the path was parallel.

The torture test printed decently on a third attempt, but the slicing profile was perhaps too conservative in terms of layer height, so the results were coarse; and the printer extruded more plastic than necessary, resulting in constrained bolt holes and runny vertical features. The printed nautilus gears worked immediately upon assembly, showing off the Jr.'s ability to print thin, stable walls.

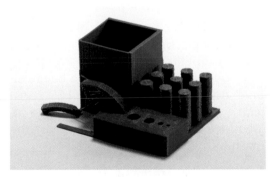

Replicator 2

- *http://makerbot.com*
- Tested by Emmanuel Mota and Eric Weinhoffer
- Written by Emmanuel Mota

Faster, quieter, and with a paper-thin 100-micron minimum layer height, the Replicator 2 is worthy of its 2.0 badge. It sports a new steel frame, oil-impregnated bronze linear bearings, and a 37% larger build volume, and it's optimized for PLA.

The Replicator 2 has a sleek modern look with an all-black powder-coated steel frame and has kept the same overall footprint and basic design while enlarging the build volume. A larger, more responsive LCD panel with a keypad on the front-right corner allows for easy control and monitoring of the machine.

A highlight is the new Cold Pause feature that pauses the print, cools the extruder, and waits for you to resume where you left off, which can come in handy.

Another fun, though perhaps less useful, function of the keypad is the ability to select

the color of the interior LED lighting to suit your mood. The build area of the Replicator 2 is now inhabited by a quick-release, frosted acrylic bed specifically made for use with PLA (polylactic acid) plastic.

Without the need to wait for a heated bed, the Replicator 2's warmup time to start a print is much shorter than MakerBot's previous ABS-printing models.

Bed leveling is easier, thanks to the new three-point bed leveling system. Instead of one adjustment screw in each of the four corners, three screws are arranged in a triangular shape, widening the center point; a twist of two screws can quickly level the bed.

The upgraded features on the machine are complemented by brand-new software. MakerBot's new MakerWare replaces ReplicatorG, with a cleaner and more intuitive user interface that lets you scale, rotate, and arrange multiple models on the build platform. Slicing is now performed by the Miracle-Grue engine within MakerWare, which is significantly faster than Skeinforge.

The Replicator 2's initial setup, out-of-box to first print, was simple and quick—it took us less than 15 minutes, including attaching the extruder and loading the filament.

Solidoodle 2

- *http://solidoodle.com*
- Written by Ethan Hartman
- Tested by Ethan Hartman and Eric Chu

Solidoodle is the company started by Sam Cervantes, an early MakerBot alumnus. Solidoodle has adopted what's become a popular standard: Sanguinololu electronics (*http://reprap.org/wiki/Sanguinololu*) and Repetier Host.

To keep the price low, it cut some corners, and for the most part, they were the right ones. The frame is a basic welded box. Rods are secured by hose clamps. A number of brackets and parts are 3D printed and ugly, though this is standard practice for RepRaps. (We'd like to see designs for these made available, just in case.) The Solidoodle isn't the prettiest, but none of this should have any functional impact: score one for frugality.

While it's doubtful that unboxing will generate the sort of loving pictorials given to new Apple products, the basics are there: assembled printer, USB cable, printed startup guide, one replacement Kapton sheet for the build surface, and a tiny supply of 1.75 mm ABS filament. (The Solidoodle will also take PLA, but ABS is recommended.)

There are a few pain points, however. You can't load or change filament with the metal enclosure in place, so grab your screwdriver, and be careful—tilt too much as you're removing the top, and you'll tweak the USB connector on the electronics sitting exposed on the back of the machine.

So how does it print? Not too bad, though a bit more tuning could go a long way. The single profile included with the Solidoodle resulted in prints that were below average for the printers we tested. Extruder temperature seemed very high, despite nominally being set to a low 190°C. The output was wavy, and overhangs dropped a bit: the owl feathers were among the worst of the bunch.

Dialing down the temperature improved things, but we could have spent a lot more time tweaking. On the upside, the machine was fast and reliable; once we figured out that the build plate temperature has to be set manually, we had no trouble. The build plate was well leveled from the factory.

Visit the websites associated with the printer, and you'll see that the community is hard at work improving the default profiles. For example, *http://solidoodletips.wordpress.com* abounds with tweaks.

Type A Series

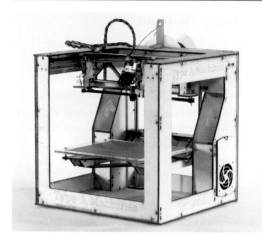

- *http://typeamachines.com*
- Written by Eric Weinhoffer
- Tested by Eric Weinhoffer and Keith Ozar

Type A Machines' Series 1 is one of the largest 3D printers we reviewed. Based in San Francisco, Type A's tiny crew of Andrew Rutter and a handful of hackers out of Noisebridge and TechShop began constructing Series 1 prototypes in August 2011. By the time Maker Faire Bay Area 2012 rolled around, the team had several iterations of their machine on view.

The Series 1 is an open hardware product, meaning you can download the pertinent case and equipment files, build your own, and make modifications at will. Like Maker-Bot's Replicator 2, the Type A Series 1 is optimized to print in PLA plastic.

The machine's 9"-cubed build volume is so big (1.2 liters) that one of the "bonus" prints we did during our review weekend was a full-scale, wearable hat. The jumbo volume is also perfect for printing multiple parts or even whole assemblies at once. The build platform is made of laser-cut acrylic, and it's held in place on the Z stage between the head of a bolt and a spring at each corner. To level it, you simply adjust these four bolts; to remove it, just pull it toward you to move larger slots over the four bolts, and lift it free.

Changing filament is a breeze: pull the lever back and slide the filament out. It's completely exposed, which leads us to believe it'll be easy to troubleshoot.

Another benefit of the Series 1's construction: speed. The Type A crew claims their frame design can clock in at printing speeds of 90 mm/sec and travel speeds of 250 mm/sec. It's also quite accurate—it will print beautifully at a layer thickness of 0.1 mm and will happily go all the way down to 0.05 mm (50 microns), where the stepping of layers is difficult to detect.

The Series 1 performed extremely well in our test prints. The large bed let us print all the parts of the nautilus gears in one go, the owl came out beautifully, and achieving a perfect snake print was no challenge. However, like many of the other machines, the Series 1 wasn't able to handle the extremely small arch in our "torture test."

The Series 1 has a few downsides. Our demo unit was loud, especially when moving at high speeds—a problem that might be minimized with grease and the tightening of bolts.

Ultimaker

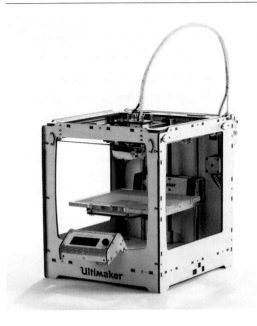

- *http://ultimaker.com*
- Written and Tested by John Abella, Eric Chu, and Matt Griffin

Ultimaker was founded in the Netherlands by one of the most active RepRap advocates and developers, Erik de Bruijn, joined by Siert Wijnia and Martijn Elserman. The first Ultimaker kits shipped in early 2011, and the company has grown to more than 20 people.

The Ultimaker was the only printer we reviewed to feature a Bowden-style extruder, where the filament drive mechanism is physically separated from the extrusion nozzle. As a result, none of the motors add to the toolhead's weight. A low-mass toolhead and stationary motors results in a lightweight gantry system, ideal for fast prints.

Updated parts, including new drive bolts and the new V2 hot-end, show a constant drive to improve the product and extend its life. The electronics have been redesigned

and updated numerous times. Our review unit also included the UltiController, an add-on that lets you adjust print speed and temperature on the fly, print from an SD card, monitor builds, and do other maintenance tasks untethered to a computer.

While the Ultimaker can print in ABS or PLA, it's engineered primarily for PLA, so there isn't a heated build platform—a near-requirement for large ABS prints.

The Ultimaker remains a great kit for buyers looking to get under the hood. The newest kits ship with the latest parts, and makers can upgrade as new parts become available.

Setup was moderately easy. The print bed's leveling process is less fussy than that of some other printers, and it tends to stay level without much intervention. Belts for the x- and y-axes are tensioned with just a hex driver and tend to stay calibrated, but there's no easy way to finely adjust the z-axis limit switch.

We achieved excellent quality prints after getting the Ultimaker dialed in—once calibrated, it's the most accurate of the DIY printers. The snake and owl came out great, but the torture test had lots of stringing, mainly due to retraction being disabled by default.

John Abella is an obsessive hobbyist and 3D printer enthusiast who has run 3D Printer Village at World Maker Faire New York since 2010. He's currently teaching 3D printer assembly workshops with BotBuilder.net.

Cliff L. Biffle is an engineer at Google and a member of Ace Monster Toys, a hackerspace in Oakland, California. He enjoys using science as a verb.

Eric Chu is a MAKE Labs Alumnus, an engineering student, yo-yo hacker, robot builder, and fried rice aficionado.

Matt Griffin is the Director of Community & Support at Adafruit Industries, a former MakerBot community manager, and author of the forthcoming MAKE book Design and Modeling for 3D Printing.

Ethan Hartman is a customer service and documentation specialist for technology companies. He worked for MakerBot from 2009 until August 2012.

Lyra Levin is a climber, aerialist, contortionist, parkour noob, and Ninja 500 rider. She is a compulsive builder of things and member of industrial arts collective Ardent Heavy Industries.

Blake Maloof is a game designer at Toys for Bob (Skylanders).

Brian Melani builds 120-pound robots in his free time. He is a MAKE engineering intern.

Emmanuel Mota is the Director of Maker Camp at Maker Media, filmmaker, photographer, and full-time geek. He got into 3D printing in 2012 when he built a RepRap from scratch.

Keith Ozar is a creative marketing professional from Brooklyn, NY. He empowers makers through special projects that highlight the potential of 3D printing.

Eric Weinhoffer is MAKE'S product development engineer for the Maker Shed. He's been an owner-operator of 3D printers since interning at MakerBot in 2009.

PART II

Software

Software for 3D Printing 3

An overview of the necessary design, slicing, and client software.

WRITTEN BY **MATT METS AND MATT GRIFFIN**

You've got a shiny new 3D printer and a brilliant idea for your first original design—now what?

Creating and printing your own unique 3D models requires three kinds of software. First, there's the 3D modeling program used to design the shape of your creation. Traditionally, the use of software to prototype physical objects has been referred to as *computer-aided design (CAD)*. Second, there's the *computer-aided manufacturing (CAM)* program (commonly referred to as a slicer) that converts your model into specific, mechanical instructions for the printer robot. Third, there's the *printer control software*, or *client*, that sends those instructions to the printer at the right time, and provides a real-time interface to the printer's functions and settings.

3D Modeling/CAD Software

Probably the most important software choice you'll make is what kind of modeling program to use. There are many to choose from, but they fall into four basic types: *solid*, *sculpting*, *parametric*, and *polygonal*. Each type will help you turn your idea into reality, but one may be handier for, say, designing a mechanical part, and another for sculpting an action figure.

Solid modeling programs mainly use a method called constructive solid geometry (CSG), or similar techniques, to define complex 3D shapes. Popular free solid modeling programs include SketchUp, Autodesk 123D, and Tinkercad (which runs entirely in your web browser and is shown in Figure 3-1; read a tutorial in Chapter 4). In a solid modeling program, simple "primitive" shapes like boxes, cylinders, and pyramids are manipulated to make more complex shapes, often using Boolean operations. For instance, a hollow box can be modeled by drawing two overlapping cubes, one slightly smaller than the other, and "subtracting" the smaller from the larger.

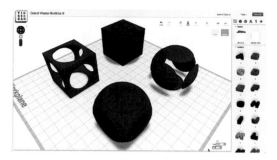

Figure 3-1. *Basic Boolean operations illustrated in Tinkercad. From back to front: union, two possible differences, and intersection of concentric cube and sphere.*

Figure 3-2. *Pixologic's Sculptris allows you to sculpt 3D models like blobs of clay. Tooltips have names like crease, inflate, smooth, pinch, and flatten.*

Solid modeling programs have three big advantages. First, the solid modeling design process tends to be more intuitive than other methods, and is often the easiest way for beginners to get started. Second, the interface usually makes it easy to set precise measurements between objects, which is handy for creating mechanical parts. Third, the software handles most issues of manifold integrity ("water-tightness") for the user automatically, despite the very large number of operations that may go into shaping a complex form.

Sculpting modeling programs, such as ZBrush, Sculptris, and Mudbox, use a more freeform interface to slice, tug, twist, and press the surface of a "blob" into the desired shape. This makes them great for forming organic surfaces such as faces or figures, but less suitable for precise parts or flat surfaces. A great tool to start with is Sculptris (Figure 3-2), little brother of the more expensive ZBrush. (Many polygonal modeler applications such as Blender, Modo, and Maya are beginning to offer built-in sculpting tools as well.)

Parametric modeling programs, such as OpenSCAD, are fairly unique; instead of drawing shapes using a mouse, objects are modeled by writing simple programs that describe how to combine different shapes together. Because each dimension can be specified precisely, this kind of tool is great for quickly creating things, such as technical parts like enclosures, gears, and other mechanical objects.

On the other hand, parametric modelers are also useful for producing generative artwork. Tools such as Marius Watz's Model-Builder and the Grasshopper editor for Rhino are geared toward generating unexpected, abstract forms by processing other objects or data, or by pure math. Designers like Nervous System use them to create complex organic shapes (Figure 3-3) that would be practically impossible to model by hand (Figure 3-3).

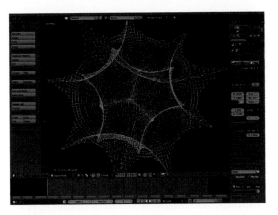

Figure 3-3. *Nervous System's "Convolution" bangle, stainless steel, based on simulated forces in a cellular network.*

Polygonal modeling programs represent objects using thousands of tiny triangles arrayed together in a mesh that defines model surfaces. Notable examples include Blender (Figure 3-4), 3ds Max, Maya, and Modo. They're great for 3D graphics and animation, but require a bit of care when used for 3D printing to make sure that meshes remain manifold or watertight (i.e., without missing polygons or disconnected vertices). If a model is not manifold, the slicer may not be able to tell its inside from its outside and may refuse to process the model at all, or may produce G-code containing serious errors.

Polygonal modeling programs offer a tremendous amount of control, but are often challenging to learn. Effective mesh modeling requires mastering a number of sometimes counterintuitive principles like working with "quads" (instead of triangles or "n-gons"), developing "edge-flow" to quickly manipulate models with operations like edge-cutting and loop-cutting, and using "subdivision" tools to automatically smooth jagged surfaces into more organic forms. Extensive tutorials covering these topics in most of the major programs can be found online. Watching a few videos demonstrating best practices early on can save you a lot of trouble as your skills develop.

Your CAD program will produce a 3D model in some file format, commonly STL. Depending on what software was used to produce it and how complex it is, your STL file may contain errors, such as holes or reversed normals, that will need to be corrected before it will print correctly. Your CAM software may detect these errors automatically, and some CAM packages—notably Slic3r—include repair routines that will try to automatically fix simple errors, but you cannot always rely on these to produce reasonable toolpaths. Models can also be repaired manually using a polygonal modeler. Another option is MeshLab, an advanced, open source STL processing and editing tool that is very powerful but may be intimidating to beginners.

One type of modeling program may be handier for, say, designing a mechanical part, and another for sculpting an action figure.

As you become more experienced with 3D printing, you might want to consider investing in a commercial STL analysis and repair tool such as Netfabb Studio. While their Basic suite works well for solving manifold issues quickly and effectively, the Professional version allows you to target specific elements of the model for manipulation, decimation, and re-meshing, as well as offering stable Boolean operations to split up a model into multiple parts. The Professional package also offers built-in slicing utilities and drivers for operating some of the printers directly, in some cases entirely replacing the CAM/client pipeline.

Slicing/CAM Software

Once you have a manifold, error-free 3D model, it must be converted into specific toolpath instructions that tell the printer where to move the hot-end, when to move it, and whether or not to extrude plastic along the way. This process is sometimes referred to as *skeining* or *slicing*. The standard format for these instructions is a simple programming language called *G-code*.

Historically, most printers have relied on the open source Skeinforge engine for preparing G-code from model files. Recently, however, alternative slicing programs have started appearing, most notably Slic3r, which has been slowly overtaking Skeinforge as the

tool of choice. For more on how to use Slic3r, see Chapter 5.

A fairly recent closed-source utility called KISSlicer, available in free and pro versions, boasts some unique features, such as adaptive sparse infill (using more material near the edges of a print and less in the center) and multi-extruder support (using different material for separate models, support structures, and infill).

Though most slicing engines can be run as standalone programs, they're commonly built into integrated printer client packages like ReplicatorG (Figure 3-5) and Pronterface, so that the same interface used to control the printer can also be used to load and slice 3D models directly.

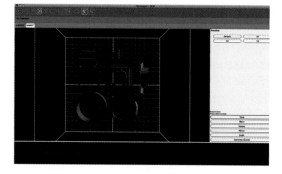

Figure 3-5. *Laying out a build plate in ReplicatorG. All of these parts will print simultaneously.*

Note that because a 3D print proceeds layer by layer, the G-code to print a single copy of a model is very different from the G-code to print, say, four copies side by side. If you want to print multiple parts per job, one option is to simply lay out build plates, as they're called, directly in your 3D modeling program. Another option, which many find more convenient, is to lay out build plates at the CAM level. Many slicing engines, as well as integrated print environments like

ReplicatorG, now provide tools that allow easy scaling, repositioning, and duplicating of CAD models before slicing. These usually include a virtual environment that shows how everything will fit into the printer's build chamber.

The slicing program will provide an interface to adjust a number of variables related to print speed and quality, such as layer height, maximum print-head speed, infill density, number of "shells" surrounding the infill in each layer, and whether or not to print support structures or "rafts." Many slicing engines have built-in profiles to get you started, and most work well right out of the box. Eventually, you'll probably want to experiment with these settings to suit specific geometry or design challenges.

A handy practice when getting familiar with slicer settings is to use a *G-code visualizer* to preview the print. A visualizer will display the G-code commands as a series of lines to represent the print-head toolpaths. Scrolling through the layers can help you learn how the slicing software tackles the geometry of the original object, and will reveal errors without using up any plastic. Saving a series of G-code "drafts" of a figure before actually running a print job is a great way to gauge the effect of adjustments to the various slicer settings. If you're using ReplicatorG, grab Pleasant 3D (for Mac, shown in Figure 3-6) or GCode Viewer for Blender (cross-platform). Both Pronterface and Repetier-Host have built-in G-code viewing utilities.

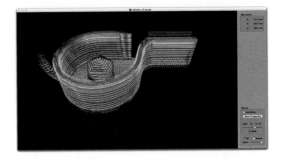

Figure 3-6. *Visualizing G-code in Pleasant3D. The interface allows you to scroll through the toolpaths one layer at a time.*

Printer Control/Client Software

Finally, there's the *client*, which is basically the printer's real-time control panel. It provides a software interface where you can start, stop, or pause the printing process at will, as well as set the temperature of the extruder nozzle and the bed heater, if present. The client will usually provide a set of directional buttons that allow you to incrementally move the print head in any direction, which can be useful for bed leveling, calibration, and manual zeroing.

Historically, many machines relied on ReplicatorG for machine control. Recently, though, some alternatives have appeared, and the amount of innovation is impressive. The Printrun suite (featuring Pronterface) and Repetier-Host are the most actively developed and used. Ultimaker has been developing the open source Cura package which is feature-packed and easy to use. Some closed-source printers, such as PP3DP's Up and MakerBot, ship with custom client software that will usually include a similar set of features.

In use, the essential function of the client is to send toolpath instructions to the printer over a WiFi or USB connection. Many printers are designed for operation in "untethered" mode, in which the printer runs on its own without a computer connection. In untethered mode, no client program is necessary; the printer automatically reads and follows CAM instructions from an SD card or USB thumb drive plugged into it directly. Untethered printing can be useful, for instance, for long-running prints during which you may want to use your computer elsewhere, or if you have more printers than computers to run them. CAM information is usually stored on removable media as G-code instructions. For a rundown of all the available printer control and slicing software, see "3D Printer Frontends" on page 199 and "Slicing Software" on page 200.

What Next?

Your printed object will inspire improvements and new ideas. The design pipeline is really a cycle.

Now that you've got the basic workflow down, you're ready to make anything! Remember that 3D design and printing is an iterative process, and that things rarely turn out perfectly the first time around. If you aren't comfortable with any of the tools that you tried, be sure to look at others—there's no reason to limit yourself to only one workflow. Experiment, tweak, observe, repeat! Try to learn something from each mistake, and always remember to have fun.

Matt Mets is a maker who uses electronics to create playful objects that teach and inspire.

Matt Griffin is the Director of Community & Support at Adafruit Industries, a former MakerBot Community Manager, and author of the forthcoming MAKE book Design and Modeling for 3D Printing.

3D Design for the Complete Beginner

4.

Use Tinkercad to design a robot-head pencil topper in minutes.

WRITTEN BY BLAKE MALOOF

Three-dimensional printing offers exciting applications, from art to product design to rolling your own replacement parts. But if you're new to this technology, the modeling software and printer hardware can be intimidating. The learning curve is steep as you go from concept to 3D model to printed object. If only you could create a simple design on your desktop, have it printed on a high-end 3D printer, and delivered to your door. Turns out, you can!

Tinkercad is an intuitive, browser-based CAD modeling application that allows you to quickly box out your design and hit a Print button to send it off to 3D printing services like Shapeways or Sculpteo. You can go from concept to ready-to-print 3D model in minutes with little to no modeling experience.

To get you started and to show just how easy it is to make something fun and unique, I'll walk you through the process for creating a robot-head pencil topper. All you need is a computer with an Internet connection.

Understanding Positive and Negative Space

To utilize Tinkercad to its full potential you need to understand the concepts of positive and negative space. For example, you can place a box, which represents a solid, positive shape, and you can also place a hole, a negative shape. Holes remove any solid material within their shape.

This will become very important when you submit your model for printing. Because the cost depends on the solid volume of the print, you'll want to make the object as hollow as you can.

1. Create a Tinkercad Account

Create a Tinkercad account (*http://tinkercad.com*). Then click the "Design a New Thing" button. Tinkercad will give your new thing a name; change it by clicking the gear icon.

2. Make a Hole

To make the hole that your pencil will fit into, measure the width and height of a pencil eraser. It's about 8 mm at the widest point, so click and drag a cylinder onto the Tinkercad "workplane" (the blue grid) and scale it to 10 mm diameter. This will provide 1 mm of wiggle room all around the pencil. To scale something in Tinkercad, first select it by clicking on it, then click on one of the small white rectangles situated around the object—these are called handles—and drag it.

To scale the shape uniformly in all directions, hold the Shift key while dragging.

The height of the eraser is about 20 mm, so stretch the cylinder to 20 mm using the top handle. Don't use the Shift key here because you want to stretch only the height, not the diameter.

Now, with the cylinder selected, click the Hole icon (next to Color in the upper-right corner of the workplane window). This will turn the cylinder into negative space, indicated by gray stripes (see Figure 4-1).

> *If you make a mistake, use the Undo button or type Ctrl-Z (Cmd-Z on a Mac) to undo the step.*
>
> *It's often useful to view your design from a different angle. To do this, click on the arrow buttons in the upper-left corner of the window. The + and – buttons are for zooming and unzooming.*

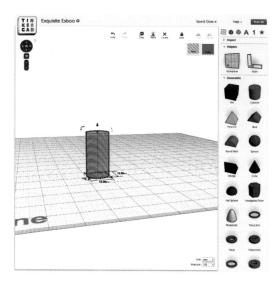

Figure 4-1. *A negative space object*

3. Make the Robot Head

Make a box for the robot head. You can make it any size, as long as the width and depth will encompass the negative cylinder with at least 1–2 mm on each side. I made a head that was 34 mm wide × 24 mm deep × 24 mm high, and placed it alongside the hole (Figure 4-2).

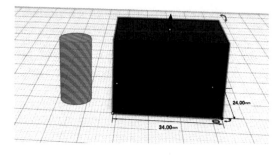

Figure 4-2. *Make the robot head*

4. Align the Head and the Hole

To make sure the hole is in the center of the box, use the handy Align tool. Select both the box and the hole cylinder by Shift-clicking each object, or click-dragging your cursor to create a Select box around them, or using Ctrl-A (Cmd-A on a Mac) on your keyboard to select everything. Look for the small gray circle with three white lines next to your selection. Click it and select Align.

Tinkercad will highlight both objects with alignment dots. Click the two middle dots on the horizontal plane to position the hole in the center of the box (Figure 4-3). Don't center the hole on the vertical axis because that would seal the hole entirely inside the box, which would make it impossible to fit a pencil in the bottom.

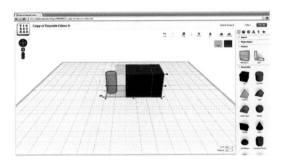

Figure 4-3. *Align the head and the hole*

5. Combine the Head and Hole into a Single Object

Select both pieces and combine them into a single object by clicking the Group icon (Figure 4-4).

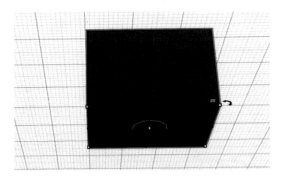

Figure 4-4. *Combine into a single object*

6. Make the Head Hollow

Make a hole box to put inside the robot's head. Size it 4 mm smaller than the head in both width and depth. This will leave the head with 2 mm-thick walls.

We don't want the inside to be completely hollow, as we need material at the bottom to hold the pencil in place. So, leave about 10 mm of material at the base by making your hole box 10 mm shorter than the head. In my case, the interior hole is 30 mm wide × 20 mm deep × 14 mm high (Figure 4-5).

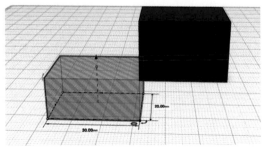

Figure 4-5. *Create an interior hole*

Use the Align tool to align the hole box to the top of the robot head, by selecting the topmost vertical dot (Figure 4-6). Now select just the hole box and use the arrow handle to move it down 2 mm (Figure 4-7).

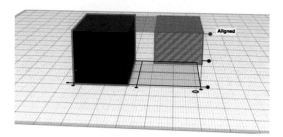

Figure 4-6. *Select the topmost vertical dot*

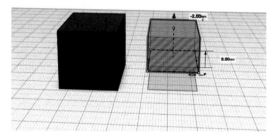

Figure 4-7. *Select the hole box and use the arrow handle to move it down 2 mm*

Select the hole box and the head box again, and align them on the horizontal plane (but again, don't click the vertical align buttons or else the hole box will move to the top or bottom of the head box).

Select all the pieces and combine them into a single object by clicking the Group icon. You now have a nice 2 mm border on the top and sides of your hollow robot head.

> *By default, the workspace has a grid snap size of 1 mm, so when adjusting the space between two objects, you can use the arrow keys to move the selected shape in 1 mm increments.*
>
> *Notice that a selected shape casts a shadow on the blue gridded plan of your workspace. Use this shadow to help you place your pieces into their proper locations.*

7. Make Your Robot's Mouth

Start by making a mouthpiece with a box that's smaller than the robot's head.

To create the speaker-grill ridges, make a series of even smaller hole boxes. You can make one and create copies by clicking the Copy button; using copy-paste keyboard commands; or clicking on the original, holding the Option key, and dragging off a clone. Space them evenly apart and slide them into the mouth box (Figure 4-8).

Then, with only the hole boxes and the mouth box selected, click Group. This will apply the hole objects to the box, cutting out the negative shapes, and turning them into holes (Figure 4-9).

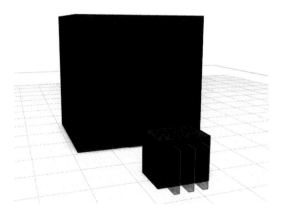

Figure 4-8. *Construct the mouth ridges*

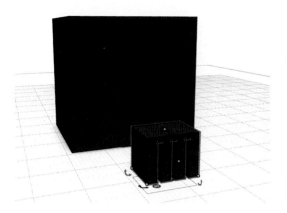

Figure 4-9. *Create the mouth ridge holes*

Place the mouthpiece so it sticks out from the front and bottom of the head box. (Don't push it in so deep that it intersects the pencil hole.) Group the head and the mouthpiece (Figure 4-10).

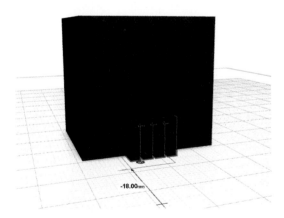

Figure 4-10. *Group the head and the mouthpiece*

8. Make Your Robot's Eyes

To make the eyes, drag two cylinders into the workspace and make them holes. See those curved arrows next to the cylinder? Click and drag them to rotate the cylinder into the proper orientation (Figures 4-11 and 4-12).

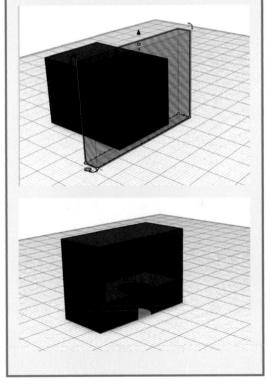

X-Ray Vision

How can you verify that the hole box is positioned correctly inside the head? Here's a little hack that provides a clearer view inside your work.

Drag a box onto the workplane, turn it into a hole, and change its dimensions until it's larger than one side of your design. Move the hole box so that it intersects your design, select all the objects, then click Group. *Voilà!* An interior view. Now, just Undo your way back to where you left off.

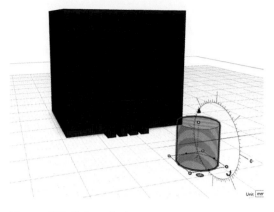

Figure 4-11. *Get a cylinder*

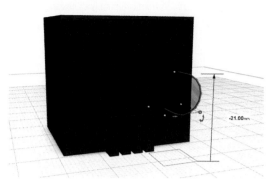

Figure 4-13. *Intersect the cylinder with the head by 1 mm*

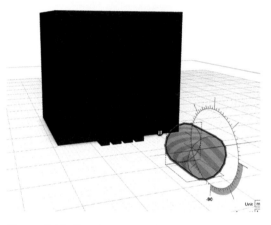

Figure 4-12. *Position it into the proper orientation for eyes*

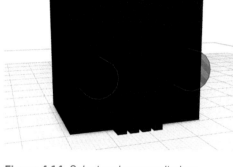

Figure 4-14. *Select and group cylinders*

You need to raise the cylinders above the workplane grid so you can place them above the robot's mouth. Use the black arrow handle above the cylinder to do this. Make the cylinders different sizes to add a little personality. Be sure the cylinders intersect with the head by 1 mm. Select the cylinders and the head and Group them (Figures 4-13, 4-14, and 4-15).

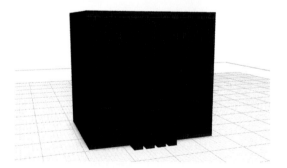

Figure 4-15. *Completed print-ready robot head pencil topper*

Tada! You now have a print-ready robot-head pencil topper (Figure 4-16). Just hit the

Print 3D button, select a printing service, and order yourself a new toy that you designed!

Figure 4-16. *Finished!*

Check out the 3D model here (*http://make zine.com/go/topperbot*).

Blake Maloof is a game designer at Toys for Bob (makers of Skylanders), and he sometimes writes about games for MAKE.

Getting Started with Slic3r

5

Slic3r is a free program that prepares STL files for printing.

WRITTEN BY ERIC WEINHOFFER

So you have a 3D printer and a 3D file, but now what? Well, you have to slice it up into layers and create a G-code file, which you'll then send to your 3D printer. There are many software options for slicing 3D models in preparation for 3D printing including: Slic3r, KISSlicer, CuraEngine, MakerBot Slicer, and Skeinforge. (See "Slicing Software" on page 200 for more on each of these options). Some of these "slicers" are integrated into printer control software and some, like Slic3r and KISSlicer can be used independently of control software.

Slic3r has become a popular option because it's open source, cross-platform, free to use, relatively quick, and extremely customizable.

I'll describe how each of the many settings relates to the actions of your 3D printer, and how to correctly adjust them to optimize your machine for your application. I don't have experience with tweaking all of these settings (there are a lot), but I'll do my best to describe what they do.

I recently read RichRap's fantastic guide, "Slic3r is Nicer" (*http://richrap.blogspot.com/2012/01/slic3r-is-nicer-part-1-settings-and.html*), and I recommend that you give it a read as well. Although Rich has a lot of nice photos and great explanations in his tutorial, it is almost a year old, and a lot has been added to Slic3r since then. Unlike me, he does cover extruder calibration in his tutorial, which is an optional, although beneficial, process.

The manufacturer of your 3D printer will most likely provide you with some default slicing settings: hopefully as an exported *.ini* file that you can import, but they might instead give you a list of numbers that you have to manually enter into Slic3r. If they supplied a *.ini* profile file, I'd recommend starting with that and tweaking settings from there (you can import a profile in Slic3r by going to File→Import Config).

Despite the fact that I provide good starter settings here, there is no set formula that will work well for all machines, so experimentation is required if you really want to optimize your prints.

You can download Slic3r for free from the Slic3r website (*http://slic3r.org*). Now open it up and let's get started!

Unless otherwise noted, photos of prints in progress are from John Abella.

Step 1: Name Your Profile

The application is broken up into four tabs: Plater, Print Settings, Filament Settings, and Printer Settings. The Plater tab is the most self-explanatory, and typically the last place you'll end up before slicing, so we'll come back to that later.

One of the neat things about Slic3r is how easy it is to create, and recall, a bunch of different profiles (Figure 5-1).

After changing any setting, clicking the Save icon will bring up a text box, where you can change the name of the profile (Figure 5-2).

Try creating a profile not only for each separate printer, but for each specific type of print as well, like "Ultimaker Hollow Part" or "Ultimaker Super Fast."

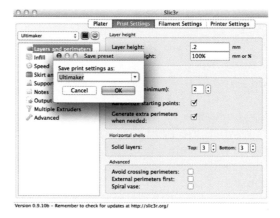

Figure 5-2. *Saving a profile*

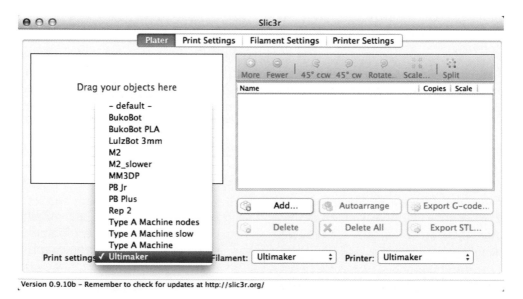

Figure 5-1. *Picking a saved profile*

Step 2: Print Settings

The first subset of Print Settings is "layers and perimeters." The layer height (Figure 5-3) is the distance the z-platform (or extruder) moves between each layer. A smaller layer height will generally result in a better looking, smoother part, but it will also take longer to print. Anywhere between 0.2 and 0.3 mm is probably a good place to start.

Many machines on the market today will handle a layer height of 100 microns (0.1 mm) without a problem.

A print with a layer height 0.1 mm will have twice as many layers as a print with a 0.2 mm layer height, and will therefore take twice as long to slice and print.

Perimeters and Solid Layers

Perimeters (or shells) are also important (Figure 5-4) in that they add to the strength of your print. A perimeter value of two specifies that the printer will draw two solid outlines around the edge of the part it's printing, on every layer (Figure 5-5). I've found that two is usually a good place to start, but three-perimeter prints are common as well.

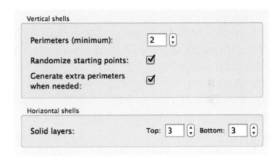

Figure 5-4. *Setting perimeters*

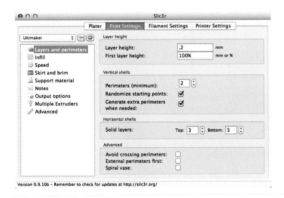

Figure 5-3. *Setting the layer height*

The first layer height specifies the height to use for the first layer you print. It can be entered in mm or % (a first layer height of 200% will be twice the standard layer height). You can use a thicker first layer to make sure your print has a stronger base for all successive layers.

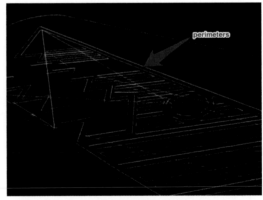

Figure 5-5. *The effect of perimeters*

Randomizing the starting point of perimeters will prevent a visual indentation from appearing on the side of your part, so I'd recommend keeping that box checked. Allowing Slic3r to generate extra perimeters when needed is also a good idea.

Solid layers are completely filled in with plastic (Figure 5-6), which is why it's usually smart to have a few of them on the bottom and top of your part. I'd recommend doing at least two solid layers on the bottom, and stick with at least one on the top.

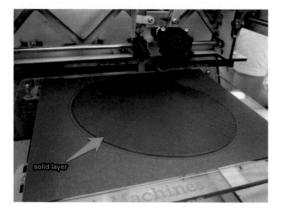

Figure 5-6. *Solid layers*

Keep in mind that if you're printing a very large part, each solid layer will take up a good chunk of time, so dial those values down if you value print time over part strength.

Infill

Fill density (Figure 5-7) is the percentage of each layer that will be filled in with plastic (0.2 = 20%). You shouldn't have to go above 60% for any reason, unless you want a really dense part. A 20% fill is just fine for your everyday prints, but adjust at will and play with the parts once they're complete to feel the difference in structural stability.

A density of 0 will only print the perimeter(s) of your part, so it will be completely hollow.

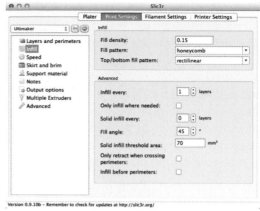

Figure 5-7. *Setting fill options*

The fill pattern (Figure 5-8) is the path that the extruder takes when doing the infill. These don't have a huge impact on the structural stability of the part. The "Top/bottom fill pattern" is the pattern used on the top and bottom solid layers.

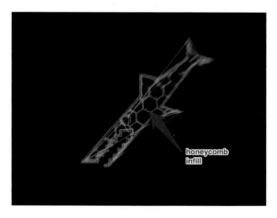

Figure 5-8. *The fill pattern inside an object*

The advanced settings give you even more control over the infill, although I don't think I've ever touched them. "Infill every 2 layers" will alternate between layers of filled (with the fill density you chose) and hollow. "Infill every 3 layers" will have two hollow layers between every filled layer, etc. I've always left this at 1, the default.

You can also choose to insert a solid layer every *n* layers, for extra stability. The fill angle is the angle at which the extruder will do its filling paths, based on the orientation of the axes in your machine. I don't see how changing this will affect your part very much, but it may have varying levels of impact based on the fill patterns you use.

I typically leave "Only retract when crossing perimeters" unchecked. We'll learn about retraction soon, and then this will make sense.

Speed

Now on to speed (Figure 5-9)! Perimeter speed determines how fast the perimeters will be printed. 50 mm/s is a good place to start, but check your printer's documentation because some printers can go much faster while others must have this value set lower. The small perimeter speed is how fast small features will be printed. This is typically slower than your normal perimeter speed, to give the plastic more time to cool down.

External perimeters are the outer perimeters of your part—the most important ones. I'd start with a speed similar to, if not exactly the same as the standard perimeter speed, and go from there.

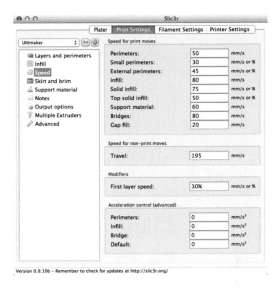

Figure 5-9. *Setting speed*

Infill speed is how fast your machine will print during the infill stage. Since clean lines and extreme accuracy aren't paramount here, crank it up! The speed I'm using here, 80 mm/s, is quite conservative, especially for the Ultimaker, but it's probably a good place to start.

Solid infill speed determines how fast the solid infill layers will be printed. These paths are more important than your everyday infills, so keep this slower than your standard infill speed. Don't bring the speed down too much, however, since 100% infill layers take a while.

The top solid infill speed is how fast the top, 100% filled layer(s) will be printed. Since it's important that these look nice, keep this speed lower than your two other infill speeds.

Bridges are used to fill in a gap, where the extruder stretches filament between two walls over air. If the gap's any greater than around 0.5", you're going to get drooping, no matter how fast you move, but moving quickly will prevent anything major. Printing material and nozzle temperature will have an effect on plastic droop during bridging. Travel speed is the speed at which your machine will move between two extrusion points. Since you'll never be extruding at this time, you might as well crank up the speed here as well. I'd recommend starting at 175 mm/s and moving up from there. Machines that use a light Bowden extruder (like the Ultimaker) can move as quickly as 300 mm/s.

First layer speed will modify how quickly your machine prints the first layer. I'd start with 50% and go from there. Read the second part of Rich's tutorial (*http://richrap.blog spot.com/2012/01/slic3r-is-nicer-part-2-filament-and.html*) on getting your first layer to stick.

The Skirt

The skirt (Figure 5-10) is an outline around the perimeter of your part, drawn by your printer before it does anything else. This is a great opportunity to "prime" your extruder, make sure your nozzle's at a good height, and kill the print before it gets too far if any adjustments are needed.

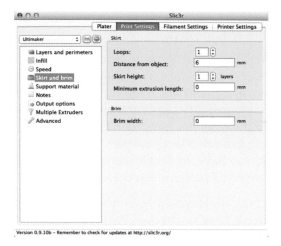

Figure 5-10. *Configuring the skirt*

If your extruder typically takes a few seconds before the plastic appears, increase the number of loops so it will make its way around your part more than once. Typically, the skirt is kept at one layer high, and anywhere from 3 to 10 mm away from the object (see Figure 5-11).

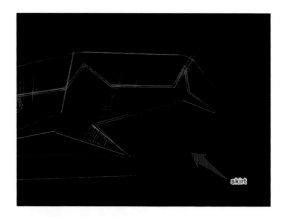

Figure 5-11. *The skirt*

Support Material

Automatically generating support material (Figure 5-12) during slicing will cause your printer to print scaffolding under overhangs and tough angles, giving you better overall results once the support's pulled away. Just check that box (Figure 5-13), and Slic3r will do all the tough work for you.

Figure 5-12. *Support material*

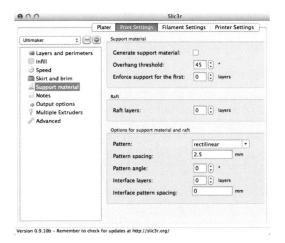

Figure 5-13. *Configuring support material*

The overhang threshold is the angle past which support will be generated. To prevent the machine from generating support for tiny protrusions that really don't need it, start with 45 degrees.

You can also select a pattern of support, just like you did with the infill, but it's probably more important here, since certain patterns are easier to break away post-print than others. Rectilinear is a good place to start.

The pattern spacing will also have a major effect on the structure of the support—a higher value will generate support that's easier to break away. The pattern angle is the angle at which the support will be printed, with respect to the x- and y-axes of your machine.

Too low of a pattern spacing will yield support that's more similar to the rest of the part, and will be hard to break away. But, too high of a pattern spacing may not provide enough support for the overhangs.

Notes and Miscellaneous Settings

Notes (Figure 5-14) are useful for your own records; they are completely optional and have no effect on the print. After printing a part and noticing how your changes affected the output, type your notes in here so you know what to change in the future.

Figure 5-14. *Taking notes*

You'll only need to mess with sequential printing if you have an automated way to remove parts from the print bed and want to print many parts in sequence. I've never bothered with changing the output options (Figure 5-15), but they're useful if you'd like to create a standard format for G-code filenames, for example.

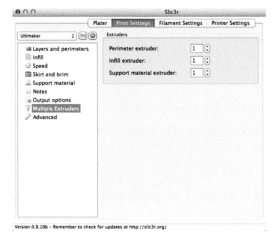

Figure 5-15. *Output options*

The multiple extruders settings (Figure 5-16) are designed for machines with just that— more than one extruder. Here you can specify specific tasks for each extruder, like support and infill.

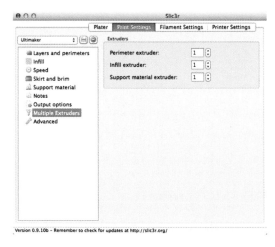

Figure 5-16. *Configuring multiple extruders*

Advanced Settings

I haven't messed with any of the advanced settings (Figure 5-17) except for the extrusion width. With accurate plastic and nozzle information (which you'll enter later), Slic3r can adjust the height of the extruder to widen the width of extrusion.

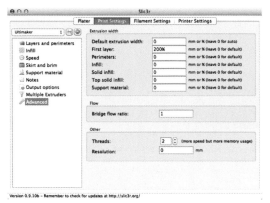

Figure 5-17. *Advanced settings*

You may want to bump the first layer width up past 100% in order to get the plastic to stick to the bed more efficiently, but I've never felt the need to adjust the other widths.

Step 3: Filament Settings

Now we're moving on to the next main tab, Filament Settings (Figure 5-18). Your machine probably came with some plastic, or you may have bought some other spools in different colors or materials. Your filament is probably advertised as being 3 mm or 1.75 mm in diameter, but that's never quite right.

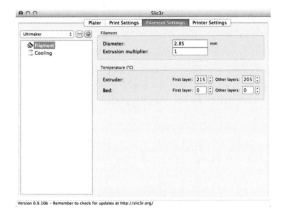

Figure 5-18. *Filament settings*

So, take a caliper or micrometer to your filament at a few different positions, and average your readings (Figure 5-19). Input the average into Slic3r.

Figure 5-19. *Caliper measurement*

The extrusion multiplier will simply alter the value you just entered into the diameter box. Unless you have a specific reason to do so, leave this at 1.

Extruder and bed temperature are also very important. You can specify a different temperature for the first layer. If anything, run your extruder hotter than usual to start, to promote extra gooeyness and stickiness.

For PLA, an extruder temperature of 185 is probably as low as you want to go (this Ultimaker profile is set for PLA printing). For ABS, I'd recommend starting at 220.

If you have a heated bed, use it at whatever temperature you feel comfortable with, since anything will help. For PLA, 60 is probably a good place to start, and 110 is good for ABS (although if your bed takes forever to get that hot, dial it down so you won't have to wait hours for a print to start).

If you don't have a heated bed, keep the bed temperature at 0. If it isn't at 0, the print will never start.

Cooling

Next up is the cooling settings page (Figure 5-20). Start with the fan settings. If your machine doesn't have a fan directed at the extruder or build platform (Figure 5-21), you can skip this step. If you have a fan, check "Enable auto cooling" and read the description that pops up when you hover your mouse over it—this setting will intelligently cool only when needed, and keep the fan off at all other times.

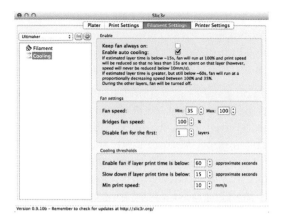

Figure 5-20. *Cooling settings*

As you adjust the following settings, refer back to the description under the enable auto cooling box to see how your edits will change the intelligent cooling activity of the machine during printing.

Fan speed is a percentage and is really up to you. Do a few prints with cooling enabled and increase the minimum fan speed if you notice that your plastic is drooping or excessively sticking to the nozzle. Bridges fan speed is how fast the fan will turn during bridging—keep this high to promote cooling and minimize drooping.

Figure 5-21. *The Ultimaker fan*

I like to disable the fan for the first layer to keep the plastic as gooey and liquid as possible, to keep it stuck to the bed (this is especially popular with PLA printing). You can also check a box to keep the fan on at all times, from print start to end.

The cooling thresholds give you more advanced control over when the fan starts. In general, layers with shorter print times (such as the tip of a cone) are more difficult for the printer to complete successfully, and therefore benefit most from additional airflow.

The thresholds to set for decreased printing speed will come with time and lots of experimentation, but I think these are a good starting point:

Enable fan if layer print time is below	60 seconds
Slow down if layer print time is below	15 seconds
Min print speed	10 mm/sec

You can set the minimum print speed fairly low; this will result in a great variation in print speed during more challenging prints.

You may find that separate cooling thresholds are necessary for different parts, so creating a different slicing profile for each may be the quickest solution—for example, one for objects with lots of narrow columns, and one for hollow objects, one for busts (where detail is important).

Step 4: Printer Settings

Now we can move onto the Printer Settings tab (Figure 5-22). Before we start with general settings, break out the ruler. Measure the usable length and width of your print area, and enter the results into the bed size boxes. The print center should be half of the bed length and width, so that the print starts at the exact center of the build platform.

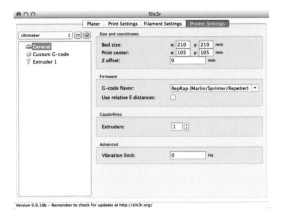

Figure 5-22. *Printer settings*

The Z offset is set at 0 mm by default, and should be left there unless you frequently change to a build platform of a different thickness. If, for example, your heated glass platform is removable, you can set the Z offset to its thickness so your machine will automatically adjust for it when you slice a part with that profile.

The G-code flavor should be fine at RepRap (Marlin/Sprinter), but definitely take a look at the drop-down and select the one that most accurately describes your machine.

Leave the "Use relative E distances" box unchecked unless you're absolutely sure that your machine uses relative positioning. Most use absolute positioning, which specifies the end point of the current move in the G-code, regardless of where you are.

The extruders value should only be changed from 1 if you have more than one extruder on your machine. If so, go back to the Multiple Extruders section of the Print Settings tab and mess with those settings.

Custom G-Code

Custom G-code (Figure 5-23) can be used to override the default calibration settings

(steps per mm) and position the extruder at a specific point at the start of a print, among other things.

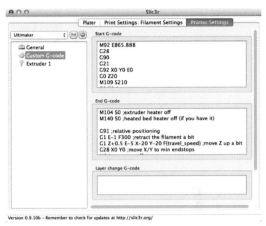

Figure 5-23. *Custom G-code settings*

The custom G-code will nearly always be specific to your model of machine, so you should check the manufacturer's documentation for these settings.

The start G-code often includes commands to zero out all three axes, heat up the extruder and heated bed, do some sort of test extrusion, and start the print.

The end G-code typically turns off the extruder and heated bed, zeros out all three axes again, and lowers the z platform for easy part removal.

Extruder Settings

Now on to the Extruder 1 section of the Printer Settings tab (Figure 5-24). The nozzle diameter should be provided by your manufacturer, but if not, take a pair of digital calipers to it and measure it yourself. It will most likely be 0.35, 0.4, or 0.5 mm.

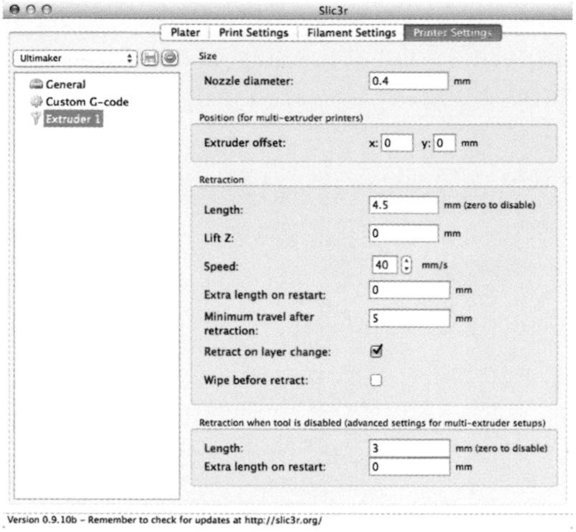

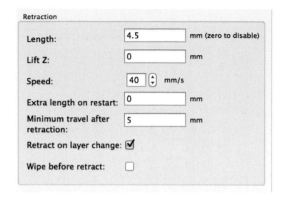

Figure 5-24. *Extruder settings*

Don't bother with extruder offset unless you have more than one extruder. If you do, this is the horizontal and vertical distance between your extruders.

Since there's so much to talk about when it comes to retraction, I'm devoting the whole next step to it.

Retraction

Retraction (Figure 5-25) is one of the coolest features in Slic3r, and will greatly improve the quality of your prints. By retracting the hot filament with the extruder motor during travel moves, plastic oozing is prevented. The length of filament to retract before moving to the next extrusion path will depend wildly on the motor and gearing you have.

Figure 5-25. *Retraction settings*

If you have no idea what to put here, I'd recommend starting with 0.75mm and moving up from there if you notice that you're still getting stringing between gaps. The value I'm using here is so high because of the Ultimaker's extruder gearing.

Lift Z will raise the extruder (or lower the bed) during retraction and before moving to the next path, where it will lower, in order to avoid knocking the part off the platform or dragging plastic with it. If you're building tall parts that may get knocked off the platform easily, set this to one layer height. If not, I would leave this at 0.

Speed is how quickly your extruder motor will reverse to retract the filament. You want this to be quick, so do some tests with your extruder and see just how fast you can retract. I'd recommend starting at 15 mm/s and building up from there, since once again, extruders will differ wildly in gearing and motor speed.

Extra length on restart is the length of plastic you'd like to extrude after traveling to a new path and prior to moving again. I don't use it, since it would just put extra plastic down where I don't necessarily need it. The only application for this may be when your extruder has serious problems starting up again after retraction, but in that case I'd just recommend dialing down your retraction length and/or speed instead of setting this to anything other than zero.

Minimum travel after retraction is the minimum distance required for the printer to retract at all between paths. If you specify 3 mm, for example, if the two paths are closer than 3 mm the extruder won't retract, to prevent the motor from doing tons of unnecessary work during an extremely complicated print. I think 2 mm is a good place to start.

The last two settings here are for multi-extruder setups. When one of the extruders is disabled, you can have it retract to prevent it from oozing while the other one is working. You can also add extra length on restart here, where it may have more use, since extruders in a multi setup are often idle for longer and may require additional priming.

Step 5: Return to the Plater

Now we can finally move back to the Plater tab! Load a part by clicking Add, or dragging it into the grid on the left. The part will automatically snap to the center of your build platform (Figure 5-26).

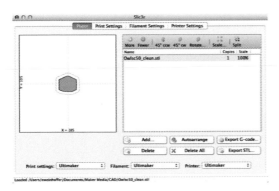

Figure 5-26. *Centered item on the plater*

You can add additional parts in the same manner, and then duplicate them by clicking "More" after selecting them (selected parts will be red). They'll be automatically arranged as you add them onto the plate (Figure 5-27).

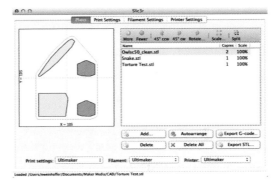

Figure 5-27. *Adding multiple parts*

You can also rotate the parts with the 45° ccw (counterclockwise), 45° cw (clockwise), and Rotate buttons. Clicking Rotate will bring up a text box (Figure 5-28) into which you can enter a specific angle. You can also scale an object with the Scale button.

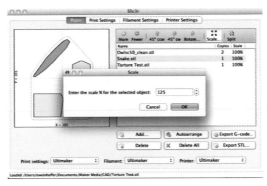

Figure 5-28. *Rotating Parts*

Working with Multiple STLs

If you import an assembly of multiple STLs, you can split it up into its separate STLs with the Split button. See Figures 5-29 and 5-30.

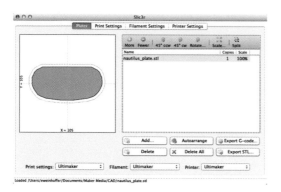

Figure 5-29. *Nautilus plate before splitting*

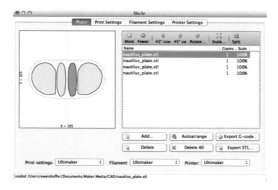

Figure 5-30. *After splitting STLs*

This is useful if you needed to print a few extra gears, but also wanted to print one single piece of another part (Figure 5-31).

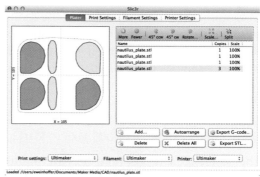

Figure 5-31. *Split STLs*

Have Fun!

That's it! Enjoy using Slic3r.

The free in-browser G-code viewer used for a few of the photos is by Jeremy Herrman: GCode Viewer (*http://jherrm.github.io/gcode-viewer/*).

Eric Weinhoffer is a Product Development Engineer at MAKE. He creates kits and sources products that we sell in the Maker Shed. Occasionally he writes about cool things for the blog and magazine.

PART III

3D Scanning

Creating and Repairing 3D Scans

Use the Kinect, ReconstructMe, and 123D Catch to capture 3D models of real world objects—then clean them up for 3D printing.

Written by **Anna Kaziunas France**

Excerpted from Getting Started with MakerBot *by Bre Pettis, Anna Kaziunas France, and Jay Shergill.*

This is all experimental. There is no "way."

— Bre Pettis

You no longer need an expensive high-end 3D scanner to create good-quality scans that are suitable for 3D printing. There are now an increasing a number of affordable ways to digitize physical objects. Some of them require additional hardware with an RGB camera and depth sensors, like a Microsoft Kinect or an ASUS Xtion shown in Figure 6-1 (see "Kinect vs. Asus Xtion" on page 61 for a comparison), but you can also use your phone or a digital camera to capture images. These images can then be converted into 3D models, cleaned up using mesh repairing software, and then printed.

Figure 6-1. *The Microsoft Kinect and ASUS Xtion*

What Is 3D Scanning?

A 3D scanner collects data from the surface of an object and creates a 3D representation of it. The Kinect and Xtion both work by beaming infrared light at an object, and measuring how far away each reflected point of light is. It then turns each individual point into a collection of points called a *point cloud* (Figure 6-2). Each point in the cloud is represented with an x, y, and z coordinate.

This point cloud is processed (or *reconstructed*) using scanning software into a digitized representation of the object known as a *mesh* (Figure 6-3). A mesh is similar to a point cloud, but instead of only using single points

Kinect vs. Asus Xtion

As soon as the community cracked open the Kinect and made it do things it wasn't intended to do, 3D scanning was one of the first items on the list. As wonderful and disruptive as the Kinect was, it wasn't the only device of its kind. In fact, other folks brought the exact same technology to market. Scanning with the Kinect is powered by hardware developed by an Israeli company, PrimeSense. PrimeSense released a software development kit (SDK) called OpenNI (Open Natural Interaction) that some people, such as the folks behind ReconstructMe (PROFACTOR GmbH), have used to develop awesome software tools for Kinect. And the great thing about this is that their software can be made to work with other hardware that uses the PrimeSense technology.

One such piece of hardware is the ASUS Xtion ($160), which has some advantages over the Kinect:

1. It is much smaller (about half the size).

2. It's lighter (half a pound).

3. It doesn't require a separate power supply (it can be powered over the USB connection).

The Xtion has some disadvantages, though:

1. It's more expensive.

2. It does not work with all software written for the Kinect.

3. It doesn't have a software-controlled motor (the Kinect has one you can use for moving the camera around).

Still, if you're looking for a portable depth camera for 3D scanning, the Xtion is well worth considering.

to form faces (flat surfaces enclosed by edges) that describe the shape of a 3D object. STL files are comprised of these triangular meshes.

123D Catch works by analyzing a group (twenty to forty) of images of an object taken from different angles. (The analysis for 123D Catch is performed in Autodesk's cloud-based systems.) By performing image analysis on the photos, 123D Catch is able to isolate an object in the photos and create a 3D mesh from the collection of photos.

Figure 6-2. *A point cloud*

(or vertices), it groups each vertex with edges (straight line segments) that combine

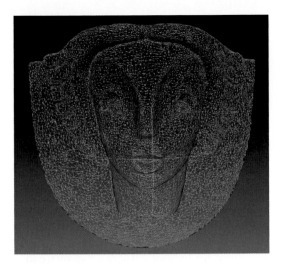

Figure 6-3. *A mesh*

Limitations

The limitations of 3D scanning depend on what technology is being used. For example, optical scanners have trouble scanning transparent or shiny objects, and digitizing probes can only scan the top surface of an object. All the software programs discussed here also have strengths and weaknesses.

What software you use to scan your model depends on the size of the model and your computer's hardware configuration. Two popular applications are 123D Catch and ReconstructMe. This chapter tells you how to use both of these. Each of these scanning packages has its own set of advantages and limitations.

In the past you needed to use a high-end scanner and expensive software, but thanks to these free programs, you no longer need to spend big bucks to get printable scans.

123D Catch

123D Catch is a free application from Autodesk that enables you to take photos and turn them into 3D models. It is available as a web-based application, an app for the iPad and the iPhone, and a desktop application for Windows. It works by taking multiple digital photos that have been shot around a stationary object and submitting those photos to a cloud-based server for processing. The cloud server uses its superior processing power to stitch together your photos into a 3D model and then sends the model back to you for editing. You can download or access 123D Catch at *http://www.123dapp.com/ catch*.

123D Catch Tips

The quality of the scan that you receive from 123D Catch is dependent on the quality and consistency of the photos you provide. Here are some general tips on how to select objects to Catch and how to plan out your Catch so that you obtain desirable results:

Objects to avoid

When choosing objects to scan using 123D Catch, avoid reflective surfaces (Figure 6-4), objects with glare, and mirrored or transparent surfaces. These objects will not work well for generating 3D models. For example, windows that are reflecting light will appear warped or bowed, like funhouse mirrors. Transparent objects like eyeglasses will appear as holes in the model.

Figure 6-4. *Avoid shiny objects—they will not Catch well*

Plan of attack

Before you start a capture project, plan out the order in which you will take your photos. It is also important to decide on a focal length. If possible, position the object that you want to Catch on a table that you can move around easily and remain equidistant from the object at all angles. Planning out how you will approach your subject is the key to success when using 123D Catch.

Mark your territory

Consider using some sort of marking system when your subject lacks discernible features or is highly symmetrical. 123D Catch has trouble with symmetrical objects, and markers will help the application to register different sides of the object. You will need four points for registration between any one image and two other images in the collection. Consider placing high-contrast tape or sticky notes on a large object. Place enough markers so that at least four are always visible from any of the positions you plan to shoot from.

Utilize background objects

When possible, utilize background objects around the object you are capturing. This will help the software parse depth. 123D Catch does not like a blank wall background with flat paint. Do not attempt to Catch objects on a flat colored surface, like a white tablecloth. You will get better results by using a background with patterns (Figure 6-5) that help the 123D Catch software clearly differentiate between the object you are attempting to capture and the surface it is resting on.

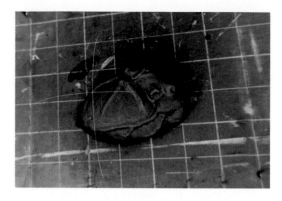

Figure 6-5. *Use a contrasting background*

What kind of camera?

Point-and-shoot cameras, like those in a regular digital camera, phone camera, or the camera in an iPad of 3 megapixels or higher will work well. We have been getting great Catches using an iPhone 4S.

Watch the Autodesk 123D Catch tutorials

Additional tips on using the 123D Catch software are available here: *http://www. 123dapp.com/howto/catch*.

Taking Photos with 123D Catch

Your first step after planning out your project is to methodically take pictures of the object you want to Catch. Here are some tips:

Provide enough information

You will need to provide enough information with your pictures for the reconstruction software to create a model. Rotate around your object, capturing a frame every 5-10 degrees (Figure 6-6). The goal is to get least 50% overlap between images. Move the camera at regular intervals and in a predictable pattern (from left to right and from top to bottom). Make sure each point in your object is appearing in at least four shots. When your photos do not have enough information, your scan may have a solid mass where there should be empty space or a gaping hole where there should be mesh.

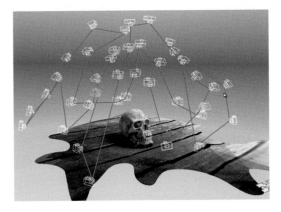

Figure 6-6. *Take photos every 5-10 degrees around the object*

Fill the frame

Try to fill the camera frame with your image (Figure 6-7). It is helpful to work consistently from high to low, and from left to right. This will help you to identify errors (should they occur) later, after the models are created. Once you start capturing frames, avoid zooming in or out. Zooming distorts your capture and may make it impossible for the application to align your set of images.

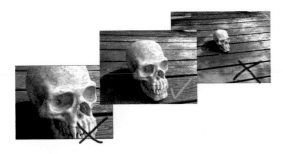

Figure 6-7. *Right way—image fills the frame*

Uniform light

Make sure there is uniform light around the thing you're trying to Catch. Avoid overexposed or underexposed images, as they hide the features you are trying to capture.

Direct light alters the exposure by creating shadows and reflective spots. The more consistent the exposure of the photographs is, the more consistent your model will be. We found that many of our best Catches were shot on overcast days or at dusk. Consider planning your outdoor Catches around these light conditions for best results.

Maintain depth of field, focus, and orientation

Blurry images will not produce accurate Catches. Review your images before leaving the scene of the Catch, and if any of them are blurry, retake the images before submitting the photos for processing. On the iPhone/iPad, you can review and retake images before submitting them for processing. If you are using a digital camera, make sure to review your images before leaving the scene of the Catch.

In addition, your images must have a consistent *depth of field*. If you are focused on the item you are Catching and the background is blurry in your photos, keep this consistent throughout the shoot. Also keep the orientation of the photos consistent. Choose either portrait or landscape and stick to it.

How many pictures?

More pictures are not necessarily better. What is important is the regular intervals and the capture of the overlapping angles of the object. Many pictures will take much longer to process, and if they are not capturing the object uniformly they will still produce poor results. The optimal number of pictures has been reported as somewhere between 20 and 55 pictures, depending on the object. If you are using the iPad or iPhone you are limited to 40 images.

Capturing detail

If you need to capture fine details, first capture the entire object at a distance that fills the frame. After you have completed a full sweep of the entire object, then move in and capture the details. Make sure that you maintain the 50% overlap between the distance photos and the detail photos, so that the software can still stitch the photos together. Be careful when transitioning from shots of the whole model to detail shots. Make sure to have transition photos that capture 50% overlap between the transitions. Do not suddenly zoom in on the detail, as this will either cause your scan to fail or produce poor results.

> By taking a whole series of close up pictures just at one level, I got really good 3D detail. Really good reproduction of very, very small depth.
>
> — Michael Curry "skimbal"

With some large objects, like statues, it may not be possible to get both very fine detail and the entire object. You may need to capture the fine detail in a separate Catch. You will need to experiment. Occasionally, we have had catches done this way completely fail on the iPhone application, and a large white X will appear after processing the Catch. Because it can take some time to see how your Catch turned out, always do one or two Catches of an object (especially if you are on a trip and may not return to it), just in case the first one fails.

If your Catch fails, consider capturing the entire object in one scan and then creating a new scan with the camera zoomed in on the fine detail.

Don't be discouraged if your first few Catches do not come out as planned; keep practicing and you will quickly get a feel for the process and how to minimize problems.

 Do not edit or crop photos before uploading! Any size, color, or tone alterations will confuse the reconstruction software and lead to less than optimal results. Upload your photos to the cloud server as they were originally taken.

Uploading Your Photos to the Cloud

Take your photos using the process outlined above and then submit them to 123D Catch via your application of choice.

If you used the iPhone or iPad application

Submit the photos via the iPhone/iPad app (Figure 6-8). The app will inform you when it has finished processing your 123D Catch scan, or *photoscene*.

Figure 6-8. *Completed photoscene on the iPhone*

If you are using the Windows desktop application

> Download your photos onto your computer. Then open 123D Catch and select "Create a New Capture." A login window will open and you will need to log into your Autodesk account (or create one).

If you are using a camera and do not have Windows

> Download your photos onto your computer. Then go to *http://apps.123dapp.com/catch* to upload them. It can take some time for the 123D Catch cloud server to process your photos, but you don't have to wait around. Your Catch photoscene will appear in the "models" section of you account when it is finished.

> *When uploading images using the Windows desktop or online application, you can select all of your images and upload them at once. You do not need to upload them one at a time.*

Downloading Your Mesh

After your photoscene is available, you need to retrieve the file in an editable format.

After the photos have been processed, and log into your account. Regardless of the method used to upload your photos, your processed scans will be present under "models & projects."

Click on a model to open it. Then download a STL file for printing at home or you can edit your model online using 123D apps.

> *The STL will often be named viewable.stl.*

There are editing tools in the online and desktop versions of 123D Catch that you can use to slice off sections to prepare models for 3D printing in both the online and desktop versions, although the online version is more up-to-date and has better tools. You can also edit your mesh using the techniques described in "Cleaning and Repairing Scans for 3D Printing" on page 70.

ReconstructMe

ReconstructMe is a 3D reconstruction system that gives you visual feedback as you scan a complete 3D model in real time. It works with the Microsoft Kinect (Xbox and PC versions) and the Asus Xtion Pro. Currently there is both a free/noncommercial version and a paid, commercial version of the software. ReconstructMe is excellent for scanning larger objects, such as people, but not great for small objects with fine detail.

ReconstructMe is currently the easiest way to get quick and complete scans, but there are a few unavoidable technical limitations that come with this type of real-time scanning. First, ReconstructMe is Windows-only, but it can be run cross-platform using virtual machines or Boot Camp on a Mac. You also need a fairly high-end graphics card to run the software. It is also picky about the version of OpenCL (a computer library that can run instructions on your graphics adapter) installed on your machine. ReconstructMe is constantly being updated, so refer to the ReconstructMe documentation (*http://recon structme.net/projects/reconstructme-console/installation*) and video card specifications.

ReconstructMe QT (http://recon structme.net/projects/reconstruct meqt/) is a graphical user interface alternative to the ReconstructMe console application. It uses the ReconstructMe SDK and is available in both free non-commercial and paid versions.

Installing ReconstructMe

Download ReconstructMe

Go to this download page (*http://recon structme.net/projects/reconstructme-console*) and download either free Lite version (*http://reconstructme.net/pric ing/*) the free developer version (*http:// reconstructme.net/projects/sdk/*).

Installing ReconstructMe on a Mac with a Virtual Machine

You can install ReconstructMe on a Mac without using Boot Camp by running Windows on a Parallels or VMware Fusion virtual machine in the same way you would install it on Windows, with one exception. You will not be able to upgrade your graphics driver for the virtual machine by downloading an update from the manufacturer. You will need to install OpenCL support separately. You can get the OpenCL CPU runtime for Windows from Intel here: *http://software.intel.com/en-us/vcsource/tools/opencl-sdk*.

If you go this route, you won't be able to perform a live scan. Instead, you'll need to first record your subject with the ReconstructMe Record tool, then complete the scan with ReconstructMe Replay.

Installing ReconstructMe on virtual machines is experimental and your mileage may vary.

Tips for Reconstructing Yourself (or Someone Else)

Once you have ReconstructMe installed, refer to the ReconstructMe website to learn how to launch the application. There are several different resolutions and modes available for scanning with ReconstructMe, and new features are being added all the time.

If you experience crashes in both the standard and highres modes, you may need to run the ReconstructMe Record tool. After saving your scan, you can play back the recording with ReconstructMe Replay and save your file as an STL.

When you have ReconstructMe up and running on your machine, here are some basic tips for scanning yourself (or someone else).

1. Sit in a spinnable office chair.

2. Position your Kinect or Xtion so that only your upper body is visible in the scan area.

3. Slowly, spin yourself around in the chair while keeping your upper body in a static position.

4. Save the file as an STL (make sure to do this after you finish your capture, while the console is open, or you will lose your scan).

5. If your graphics card or memory constraints are causing the program to crash, try using the Record feature to record your scan and then playback to reconstruct the mesh.

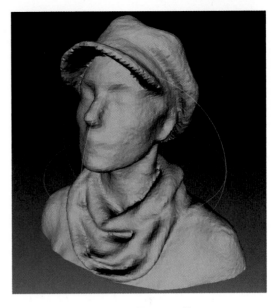

Figure 6-9. *Scan of Anna Kaziunas France*

When scanning yourself, sit with your back to the Kinect/Xtion with your computer in front of you. That way your arm movements will not be captured when you press the keys on your computer to start and end the scan.

After saving the STL, open it up in MeshLab or Pleasant3D and take a good look at it. Figure 6-9 shows a scan of Anna.

All the ReconstructMe scans in this chapter were done using Boot Camp on a mid-2010, 15-inch Mac-Book Pro running OS X 10.8.1 (Mountain Lion) with a 2.66 GHz Intel Core i7 processor and 8 GB of memory. The graphics card used was an NVIDIA GeForce GT 330M 512 MB. With this configuration, we were able to run real-time reconstruction mode but unable to run the high-resolution setting for ReconstructMe. The lack of definition in the facial features in the scan reflects these constraints. However, ReconstructMe is excellent at capturing folds in fabric, so Anna wore a hat and scarf during the scan to make up for the lack of facial definition. Other smooth fabric items, like shirt collars, ties, and smooth hair, are also captured well.

Get a Handle On It!

If you are scanning other people or things, a Kinect handle (like this (*http://www.thingi verse.com/thing:18125*)) or a Kinect tripod mount (or such as this (*http://www.thingi verse.com/thing:6930*)) can come in handy (see Figure 6-10).

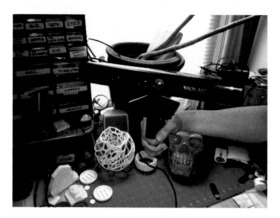

Figure 6-10. *Kinect on a handle*

Cleaning and Repairing Scans for 3D Printing

While it is becoming easier to create high-quality scans, creating valid input files is sometimes difficult. Before you can print your 3D scans, you need to clean up, edit, and repair the files to make them printable.

The most common problems with 3D scans are:

- Holes
- Disconnected parts
- "Junk" from the environment around the model or used to map the object in space that is not part of the model

- Open objects with faces that are not closed

However, analyzing STL files for errors and buildability has never been easier. Each of the following software packages has strengths that when used together, can make it easy to edit and print great looking scans.

Tony Buser created the seminal video tutorial on cleaning and repairing 3D scans that deeply informed this chapter (you can watch here (http://www.vimeo.com/ 38764290)).

Tony had also created an new updated video with a streamlined scanning cleanup workflow (here (https://plus.google.com/ 101036414115172779753/posts/ TFbYJ3Ldv44)).

netfabb

netfabb (Figure 6-11) enables you to view and edit meshes and provides excellent repair and analysis capabilities for your STL files. netfabb makes it easy to slice off bits of jagged scans and quickly repair those scans. In most cases, you will want to slice off the bottom of your model to create a flat surface against the build platform.

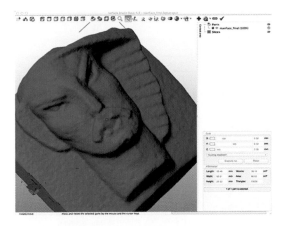

Figure 6-11. *123D Catch scan of a stone face, shown in netfabb*

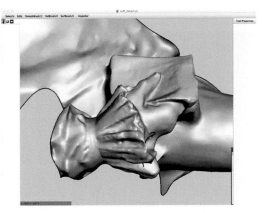

Figure 6-12. *123D Catch scan of a statue, shown in MeshMixer*

netfabb (*http://www.netfabb.com*) is available as a desktop application and a cloud service. It is also available as an STL viewer with connection to the cloud service on the iPhone. netfabb Studio is available in both Professional and (free) Basic editions. It runs on Windows, Linux, or Mac.

Autodesk MeshMixer

MeshMixer (*http://www.meshmixer.com/*) is great for mashing up individual meshes together into a new model (Figure 6-12). It works well for smoothing out bumps, blobs, and other strange artifacts that can show up in scanned files. It is also an excellent tool for capping models that are missing a side/top/bottom to make them manifold.

MeshLab

MeshLab (*http://meshlab.sourceforge.net*) can repair and edit meshes, but its Poisson filter is great for smoothing surfaces to clean up scans for printing (Figure 6-13). It's easy to rotate meshes with the mouse, so it also is an excellent STL viewer. It is available as a cross-platform desktop application and as a model viewer for iOS and Android. See also "Smoothing Out the Surface of Meshes" on page 75.

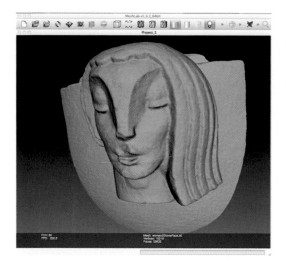

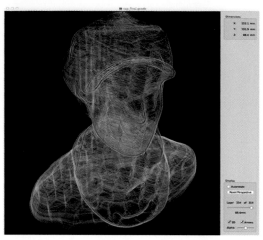

Figure 6-14. *G-code visualization in Pleasant3D*

Figure 6-13. *123D Catch scan of a stone face, shown in MeshLab*

Pleasant3D

Pleasant3D (*http://www.pleasantsoft ware.com/developer/pleasant3d/*), shown in Figure 6-14, is a great Mac-only application for previewing and resizing STL files by specified units (as opposed to scaling in Maker-Ware). It can also convert ASCII STL files into binary STL. It shows G-code visualizations, which let you preview how your model will print.

Repairing Most Scans

Most scans you create using these software programs will have a mesh that is mostly complete. However, these scans will usually have holes, junk, and other issues that you will need to fix. If your scan is missing large areas of mesh and has huge gaping holes, or is just the front or relief of a building or sculpture, see "Repairing Relief Scans by Capping" on page 77. To find out how to repair more minor issues, read on.

Repair and Clean Up in netfabb

Open netfabb Studio Basic and open the STL file of the model with Project→Open. (See Figure 6-15.)

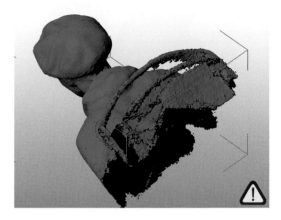

Figure 6-15. *ReconstructMe scan opened in netfabb*

To repair and clean up the model, follow these steps:

Show the platform

To help you see your model's orientation, select View→Show Platform. If you can't see the yellow platform, then you may need to zoom out.

Reorient the model

To move the part to the origin of the platform, select Part→Move, then select the To Origin button from the dialog box and click Move.

Now zoom in on your model by selecting View→Zoom To→All Parts.

Click the selection tool (the arrow). Click the model to select it, then move the selection tool over the green corner that appears around the selected model.

When the selection tool is over the green corner bracket, it will appear as a rotation symbol. Rotate the model, tilting it so the head is pointing up and the body is pointing down towards the platform, as shown in Figure 6-16.

To pan in netfabb, hold down Alt and drag the mouse.

Tweak the model alignment

You will have to change your view and rotate the model several times in order to orient it on the platform. Try to place the model so that the shoulders are at equal height.

You can change your view from the View menu or click on the cube faces in the main toolbar at the top of the screen. Align the model within the box relative to the platform.

Make sure to tilt the head back using the rotate tools to help with the overhang that can develop under a person's chin. (Remember the 45 degree rule from "Generating STL files".)

Figure 6-16. *Scan reoriented*

Watch Your Underhang

The model shown here has a severe chin *underhang*. Try to minimize this when creating your scan. The back of the head where the hat juts out also has a severe overhang. We used external support material to help print the hat back. The support material did not build under the brim and chin, but it still printed pretty well!

Slice off the jagged bits

Use the Cuts tools on the right of the screen to cut a flat bottom for the model. Drag the Z slider so the bottom of the blue cut line is cutting off the scan's jagged edges. Click the Execute Cut button, then click Cut.

You can then click on the part of the cut model that you want to remove. The selected part will turn green, as shown in Figure 6-17.

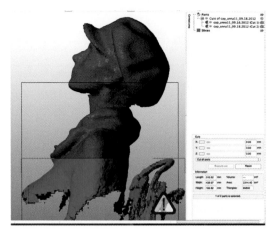

Figure 6-17. *Selected section to remove*

Remove the jagged section

Go to the Parts section that is now displayed in the top-right corner of the screen. Click the X next to the part you want to remove (the jagged parts of the scan) to delete that part. netfabb will ask you if you *really* want to remove it. Click OK.

You will now have a nice clean edge at the bottom of your model.

Move to origin

Move the part to the platform by selecting Part→Move, then selecting the To Origin button from the dialog box and clicking Move (see Figure 6-18).

Figure 6-18. *The part moved to the platform*

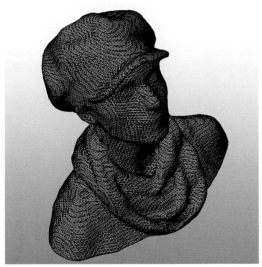

Figure 6-19. *Look at all those triangles!*

Repair the holes

Next, we will repair the holes in the model. There is probably a large hole under the chin where the scanner could not gather information, and possibly a hole in the top of the head.

Select the Repair tool (it looks like a red cross). The model will turn blue and you can see the triangles in the mesh. There will be yellow spots where repairs are necessary. To repair the model, click the Automatic Repair button.

Then select Default Repair from the dialog box and press the Execute button. netfabb will ask you if you want to remove the old part. Click Yes.

Then click the Apply Repair button at the bottom of the righthand side of the screen. In the dialog box that opens, select Yes when asked if you want to remove the old part. Figure 6-19 shows the repaired model.

Save in netfabb file format

Save your netfabb project, so you can edit it later (Project→Save As).

Export to STL

Export it as an STL file using Part→Export Part→STL.

netfabb may warn you that there are issues with your file. If you see a big red X when you attempt to export your model, click the Repair button on the dialog box. The X will become a green checkmark.

Then click Export to save the STL.

Smoothing Out the Surface of Meshes

Sometimes you want a smooth surface on a model so it makes a smooth, shiny print. MeshLab's Poisson filter will smooth out your mesh nicely.

If you are able to create a high-resolution scan from ReconstructMe, you may want to smooth it out a little for printing. For regular resolution, ReconstructMe scans, skip this step to keep more detail in the model.

Open MeshLab

Create a new project using File→New Empty Project.

Then select File→Import Mesh to open your STL file in MeshLab.

When the dialog box pops up that asks you if you want to Unify Duplicated Vertices, click OK.

Turn on layers

Go to the View menu in the top toolbar and select Show Layer Dialog.

Apply the Poisson filter

Select Filters→Point Set→Poisson Filter→Surface Reconstruction Poisson. In the dialog box, set Ochre Depth to 11 (the higher number, the better; "good" is 11). If you go any higher than 11, MeshLab may crash.

Click Apply.

Hide the original mesh

After applying the Poisson filter, there will be two layers: the original (Figure 6-20), and one labeled Poisson Mesh.

Click on the green "eye" icon by the file name to hide the original file and keep the Poisson mesh. You will see visible smoothing on the surface of the model, as shown in Figure 6-21.

Save as an STL file

Go to File→Export Mesh As. Use the default export options.

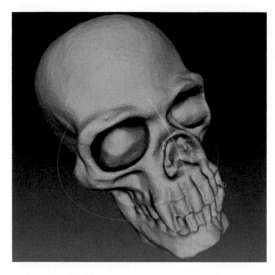

Figure 6-20. *Default mesh*

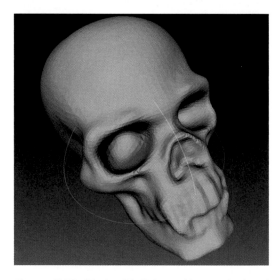

Figure 6-21. *Mesh with Poisson filter applied and original mesh hidden*

You can see a print of a skull in Figure 6-22.

Figure 6-22. *Print of the scanned skull at World Maker Faire New York*

Removing Bumps and Blobs with MeshMixer

Depending on how your scan came out after repairing and (optionally) smoothing, you may want to remove some bumps or blobs. If your model does not need any additional smoothing, you can skip this step.

Import your STL

Open up MeshMixer and import your STL file by clicking Import in the top toolbar.

Smooth it out

Select Sculpt → Brushes and one of the "Smoothing Options" from the lefthand navigation bar. Use the sliders to adjust the brush size, strength, depth, and other features.

Click and drag on the bumpy/lumpy areas to smooth them out. When you are satisfied with the appearance, export the file as an STL.

Final Cleanup/Repair in netfabb

Open the STL back up in netfabb.

If you used the Poisson filter in MeshLab, the formerly smooth bottom of your model will be bumpy. To fix this we need to recut the

bottom of our model to make it flat. Reslice off the bottom.

Repair the model and export as an STL.

Print Your Model

Your scan is now cleaned, repaired, and ready to print! Open it in MakerWare, then resize or rotate it if necessary and print it out. See Figure 6-23.

Figure 6-23. *Final printed ReconstructMe scan*

Repairing Relief Scans by Capping

Sometimes you have a mesh of a building or sculptural relief that is missing a side, top or back and you need to create a closed model by "capping" the object so you can print it. Meshes generated from 123D Catch scans often have these issues when you are only able to scan the front part of a large object. MeshMixer and netfabb can easily help you fix this problem, as well as filling minor holes or removing disconnected parts.

If your model has a lot of extra "junk" in it, it is a good idea to slice it off in netfabb before editing it in

---

MeshMixer. However, sometimes little parts will be impossible to slice off. When this occurs, you can use the lasso tool in MeshMixer to select and delete those stray bits of mesh.

MeshMixer doesn't have any labeled controls for panning and zooming around your model—you need to hold down the key combinations while dragging with your mouse/trackpad to change your view (see also *http://www.meshmixer.com/help/index.html*):

Basic MeshMixer view controls include:

- Alt + left-click: orbit camera around object
- Alt + right-click: zoom camera
- Alt + Shift + left-click: pan camera

Fixing Holes, Non-Manifold Areas, and Disconnected Components

When you have a scan that is missing large portions of the mesh, you first need to address the holes, nonmanifold areas, and disconnected components. We will attack each problem in turn.

Open MeshMixer and import your STL or OBJ file (Figure 6-24).

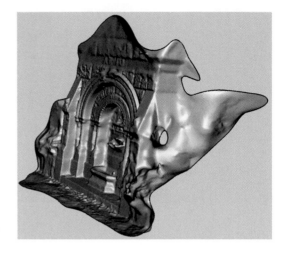

Figure 6-24. *Scan opened in MeshMixer*

In the side navigation bar, click "Analysis". Your model will now have a number of colored spheres attached to it, as shown in Figure 6-25:

- Red spheres represent non-manifold areas.
- Magenta spheres represent disconnected components.
- Blue spheres represent holes.

  Find the sphere that indicates the large hole

  Orbit around your model (Alt + left-click, drag your mouse) to identify which blue sphere is directly on the blue outlined edge that represents the large hole in the model that we want to cap. In the case of the model shown, we want to cap the back of the fountain.

  Take note of this particular sphere and make sure that you edit it last. In the case of the fountain model, the sphere that indicates the large hole is circled in the screenshot (see (Figure 6-26). You want to close all of the other minor holes first. Leave the circled sphere for last, we will get to it later when we cap the back of the model.

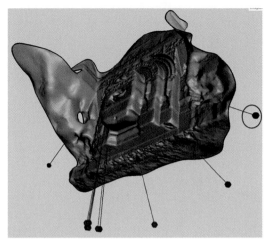

Figure 6-26. *Model shown with sphere indicating large area of open mesh*

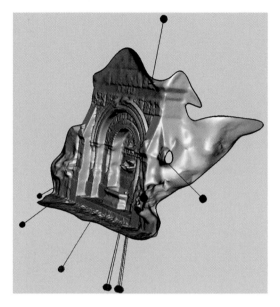

Figure 6-25. *Model shown with spheres indicating mesh problems*

 When repairing meshes with large holes or missing areas, do not click AutoRepair All as this can cause the program to immediately crash. In addition, you want to close the back of the model ourselves to control how it is closed. You don't want a big autorepaired blob, you want a nice, smooth cap.

Repair the problem areas

Clicking on a sphere will repair the problem. Right-clicking on the sphere will select the area and allow you to edit the selected part of the mesh. When you right-click, editing options will appear on the side of the screen.

First, left-click on any red or magenta spheres to close the nonmanifold areas and reconnect the components. The sphere and indicator line will disappear after you click on it, indicating that the problem is resolved.

Next, close all of the holes by clicking on the blue spheres, with the exception of the sphere that represents the large area of missing/open mesh. Orbit around the model to make sure you get them all.

Select the last sphere

Next, right-click on the last blue sphere that represents the large area of open mesh (see Figure 6-27). The blue edges will now have a dark orange tint to them where the mesh is selected, as shown in Figure 6-28.

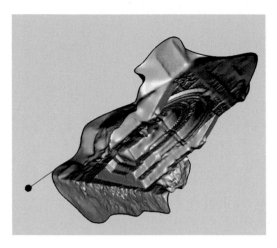

Figure 6-27. *One sphere left—time to cap the hole*

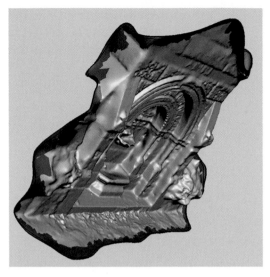

Figure 6-29. *Smoothed edges*

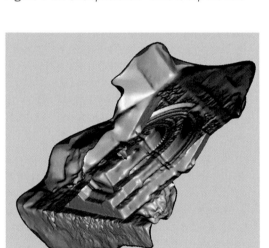

Figure 6-28. *Selected edges*

Smooth out the edges

From the menu at the top of the screen, select Analysis and then Smooth Boundary, as shown in Figure 6-29.

Then click Accept from the top menu. Figure 6-30 shows the result.

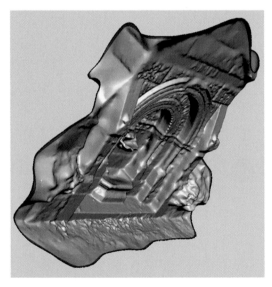

Figure 6-30. *Done smoothing*

Closing Large Areas of Missing Mesh

First repair the model and smooth the boundary, as outlined previously. Then:

Rotate the model (if necessary)

For this example, the model needed to be rotated so that the sides could be extruded (see Figure 6-31). You may need to rotate your model to get a better view of the missing mesh area.

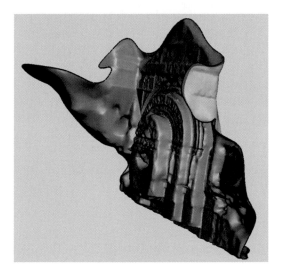

Figure 6-31. *Rotated model*

Select Extrude

With the boundary still selected, click the Select menu → "Edit" → "Extrude".

Mesh selections in MeshMixer will stay selected until you manually click Clear Selection/in the Select menu, or press Esc.

Extrude the model

From the extrusion options panel, choose Flat from the EndType drop-down menu.

Under Offset, choose an offset number that is negative. You can drag the grey bar behind the Offset label to the left or right to change the offset extrusion.

You may also need to change the Direction option to get a straight extrusion. In this example, the direction was changed to "Y Axis."

When you are satisfied, click Accept the top navigation bar. Our model now looks like Figure 6-32.

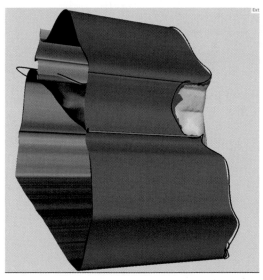

Figure 6-32. *Extruded sides*

Smooth, then rotate

From the top navigation bar, click on Modify Selection menu and select Smooth Boundary.

Then click Accept.

Rotate your model so that you are looking head on at the open area. (See Figure 6-33.)

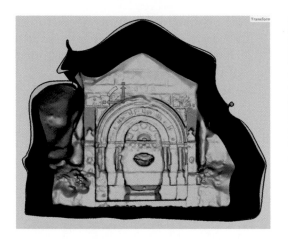

Figure 6-33. *Rotated model*

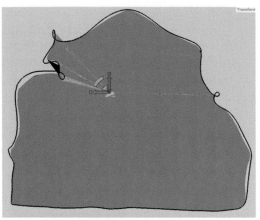

Figure 6-34. *Closing faces*

 Sometimes MeshMixer will crash at this stage. It will usually allow you to open the model again. Save often in the default MeshMixer .mix format to ensure that none of your changes are lost.

Transform faces

From the Deform menu, select Transform Faces.

Arrows in the *x,y,z* plane will appear. Scale the extrusion in by dragging on the white box between the arrows. Do not close the hole completely. Figure 6-34 shows this.

Then click Accept.

Erase and fill

Now we need to close the hole. From the Edit menu (under Select), select Erase & Fill.

Then click Accept. The result is shown in Figure 6-35.

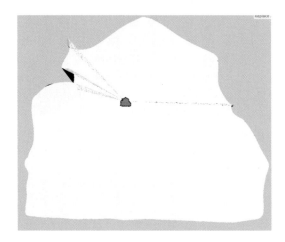

Figure 6-35. *Erasing and filling*

The model should be manifold and appear "capped" with a flat back.

Deselect

This model has some ridges in the roofline resulting from the extrusion that need to be smoothed out and repaired.

Click on the Select menu in the navigation bar and click Clear Selection, to clear the previous selection. See Figure 6-36.

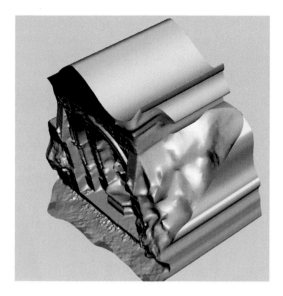

Figure 6-36. *Preparing to reduce a ridge*

Smooth with the Flatten and Reduce Brushes
These ridges will not smooth out with the Smooth brush, so we need to use additional brush tools.

Select and use the Flatten and Reduce brushes (Figure 6-37) to flatten out the bumps.

Then use the brush to soften it out. Figure 6-38 shows the results.

Figure 6-37. *MeshMixer Brushes*

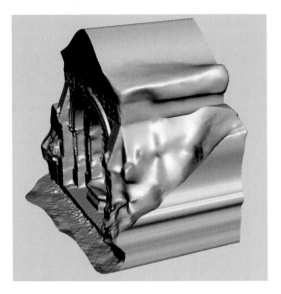

Figure 6-38. *Reduce brush results*

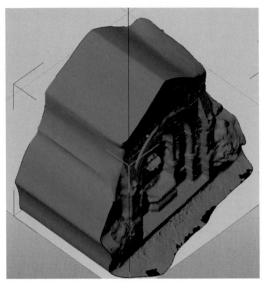

Figure 6-39. *The model in netfabb*

Export your file as an STL and then open it up in NetFabb. See Figure 6-39.

Slice and repair in netfabb

Use the same process detailed in "Repair and Clean Up in netfabb" on page 72 to slice off unwanted parts of the model, repair the mesh, and then export it as a binary STL. Your model should now be capped, cropped, and ready for printing (Figure 6-40)!

Figure 6-41 shows a photo of the final printed model.

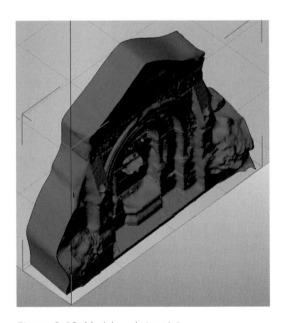

Figure 6-40. *Model ready to print*

Figure 6-41. *Printed scan of the gothic fountain outside the Providence Athenaeum*

Scan Your World

With the tools and techniques outlined in this chapter, you're ready to scan anything that you can convince to sit still for a while. And even if you end up with messy meshes, you can clean your can up well enough that you should be able to print almost anything you can scan. It's time to digitize the world around you.

The scans and models from this chapter are available on Thingi-verse (http://www.thingi verse.com/akaziuna) and 123D Gallery (http://www. 123dapp.com/Search/Index.cfm? keyword=anna+kaziunas +france).

Anna Kaziunas France is the Digital Fabrica-tion Editor of Maker Media. She's also the Dean of Students for the Global Fab Academy pro-gram and the co-author of Getting Started with MakerBot. *Formerly, she taught the "How to Make Almost Anything" rapid prototyping course in digital fabrication at the Providence Fab Academy. Learn more about her at* her website *(http://kaziunas.com) and check out her things at* her Thingverse page *(http://thin giverse.com/akaziuna).*

Print Your Head in 3D! | 7

Use digital photos and a 3D printer to make a mini plastic replica of your noggin.

WRITTEN BY **KEITH HAMMOND**

Here's a great project to get you started in 3D printing—create a 3D model of your own head and then print it out in solid plastic (Figure 7-1).

Figure 7-1. *A build plate full of heads*

A 3D printer makes an object by squirting out a tiny filament of hot plastic, adding one layer at a time. Because it adds material rather than cutting away at it, 3D printing is called *additive manufacturing*. You send the printer a file that's a 3D model of something —an iPod case, a bike part, your head—then it prints out the object for you. These machines are becoming affordable for schools, labs, libraries, and families, and there's lots of software out there for creating 3D files to print.

We chose Autodesk 123D software because it's free, it's web-based so you can use it from any computer, and amazingly, it lets you create a 3D model directly from digital photos. That way, you can do it all from home, and you don't have to get yourself scanned by a laser scanner or fiddle with a Kinect.

When you're done making your 3D model, you can print it even if you don't have a printer: you can take it to a makerspace where they have a 3D printer, or you can send it out to a service and they'll print it and mail it right to your home. We printed our heads on an Ultimaker printer, using Cura as the printer software. It was easy!

Imagine what else you could 3D print with these tools. Instead of printing your head, why not replicas of buildings or sculptures at an art museum? Or you could make models of your pets, your car—almost anything you can capture in photos.

1. Register with Autodesk 123D

Go to *http://123dapp.com* and create a free account. For this project, we'll use the web app for 123D Catch. It stitches your digital photos together into a 3D model.

Autodesk recently updated 123D Catch so you can 3D print your head two ways: either send your model out to be printed for you, or download it so you can print it yourself. (There's a powerful desktop PC version of 123D Catch, but you won't need it for this project.)

2. Take Digital Photos of Your Head

You'll want a friend's help with this part. You can use a cellphone camera or a nice DSLR—the better the camera, the better 123D Catch will work. Shooting in full shade works best.

Sit still while your friend snaps 30 to 40 photos of your head, in two separate loops moving completely around you—one lower loop, and one higher loop where the top of your head is seen clearly. This will prevent unwanted holes in your head where the software is missing part of the scene. For best results, make sure your head fills most of the frame.

If you're going to stick out your tongue or make a face, ask your friend to work fast so you can hold your expression. But remember to keep the camera still and focused when snapping each photo because blurry images may confuse the software and cause weird horns on your head.

3. Create a New Capture

In 123D Catch, upload all of your head photos. In the Model Resolution pull-down menu, select High (For Fabrication). Give your model a name and click Create Model.

Autodesk's servers will automatically stitch all your photos together to make a 3D model, and then put the model in your My Projects section.

4. Open Your 3D Model

You're looking at yourself as a 3D model! It's got a realistic texture, like your original photos. You can Dolly, Pan, and Orbit to move your view around, by using those three buttons on the righthand toolbar.

On the same toolbar, select Material & Outlines to see the 3D mesh that's underneath the texture. Cool!

5. Edit Your 3D Model

My 3D model had a crazy horn on the back of my head, maybe because we took some photos that were blurry or too far away. It also captured background elements

(Figure 7-2) that we didn't want to print. To remove major unwanted features, use the Select Faces tool to highlight them, and then Delete them. (Or highlight your model, click Invert Selection, and delete everything but your model.)

Figure 7-2. *Background elements*

To snip a horn (Figure 7-3) from your head, use the Delete & Fill tool, then use the Smooth Brush to round it off (Figure 7-4). Trim your model to size and save it under a new name.

Figure 7-3. *Finding a horn*

Figure 7-5. *The view from the bottom*

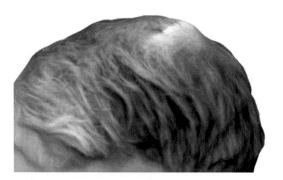

Figure 7-4. *Trimming down the horn*

6. Make It "Watertight"

Click on Inspect Model and Cap All to automatically repair any holes.

The bottom of your model will be closed now, but it might be an extended blob. For best results on the 3D printer, your model should be flat on the bottom (Figure 7-5). Click on Plane Cut Model, then drag and/or rotate the plane to where you want to slice the bottom off your model (Figure 7-6). Click Apply and your model will have a flat bottom. Resave your model to My Projects.

Figure 7-6. *Slicing off the bottom*

You can export your model as an STL file for printing now, or fool with it some more using MeshMixer software as shown in Step 7.

7. Embellish It (Optional)

MeshMixer (free from *http://meshmixer.com*) is a powerful tool for editing 3D models and merging them together. Autodesk recently acquired it, and it's frequently updated. Before using it, we recommend that you watch the video tutorials at *http://youtube.com/user/meshmixer*.

For a quick-and-dirty pedestal, open your STL file in MeshMixer. Select the whole mesh (Ctrl-A or Cmd-A), then select Edits→Plane Cut to slice off the bottom. Select the bottom face of the model and click Edits→Extrude. In the Tool Properties bar on the right, set the EndType to Flat. Then click and drag the Offset bar to extend your model, creating a simple pedestal that's perfectly flat. Click Accept and save a new STL file.

To merge your head with a fancy pedestal, start with a 123D mesh that's still open on the bottom. Select the whole mesh, choose Edit→Convert to Part, and click Accept. Look at the Parts bar on the left: your head is now a "part" you can merge with other parts. Now import an STL file of a pedestal—I like the pawn from Mark Durbin's Column Chess Set (*http://thingiverse.com/thing:19659*). Open it in MeshMixer, scale it to match your head, then drag your head onto it to merge the two. If it doesn't work the first time, try Edit→Remesh. Save the result as a new STL file.

You can do lots more with MeshMixer (Figure 7-7). Put bunny ears on your head, or stick octopus tentacles on it, or make yourself a two-headed monster. Or put your head on a Pez candy dispenser!

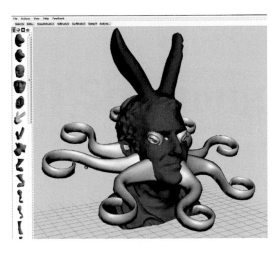

Figure 7-7. *Keithtopus*

8. Share Your Model (Optional)

When your model is done, click Publish to Gallery. Now anyone can open it in a web browser and play with it. (If you're using the desktop version of 123D Catch, you can make a video animation and send it straight to YouTube.)

9. Save Your Final Model as a Printable File (Optional)

To print your own head, you need a copy of your 3D model in a format that 3D printers can understand. Export your model from 123D Catch (or from MeshMixer) as an STL file.

If you're sending your head out to be printed by Autodesk, you can skip this step.

10. 3D Print Your Head!

We printed our heads on the Ultimaker in the MAKE Labs, which we like because it's fast and accurate—and because you can buy it as a kit and build it yourself. We've also had good success with a MakerBot Thing-O-Matic (Figure 7-8).

Figure 7-8. *Printing your head*

First, open your STL file in the 3D printer's software, which tells the printer exactly where to make trails with the hot plastic to build up your object. For example, if your MakerBot uses ReplicatorG software, import your STL file, center the model and put it on the platform, then scale it to your desired size. Next, choose Generate GCode, select the default print profile, and check the Use Print-O-Matic checkbox. Now hit Print.

Watch in amazement as your head materializes before your eyes.

If there's no 3D printer close by, that's OK—lots of service companies will print out your 3D model for you. At *http://123dapp.com*, select your project and click Fabricate→3D Print to send your file to Autodesk's digital fabrication service and receive your 3D-printed plastic head in your mailbox. It only costs about $10 for a 3"-tall head.

Or try sending your file to Shapeways (*http://shapeways.com*) or Ponoko (*http://ponoko.com*), or in Europe, try Sculpteo or i.materalise. Some of these services will even print out your head in ceramic, glass, steel, silver, gold, or titanium!

Keith Hammond is Projects Editor of MAKE *magazine. He grew up reading* Scientific American, National Geographic, *and* Spy *in the 1980s, coedited* The Nose *and* Mother Jones Online *in the 1990s, lobbied Congress for wilderness protection in the 2000s, and joined Maker Media in 2007.*

PART IV

Materials

Plastics for 3D Printing

8

An overview of 3D printing filament—from rigid to rubbery to dissolvable.

WRITTEN BY **SEAN RAGAN AND MATT STULTZ**

Desktop 3D printing filaments used to be limited to ABS and PLA, but there are now a range of different materials on the market. Basic printing temperature ranges are listed here, but keep in mind that recommended nozzle and bed temperatures vary with filament suppliers and the printer used. In addition, when printing at accelerated speeds, the upper temperature range is recommended to keep the filament moving and avoid clogged nozzles.

For a list of vendors that sell 3D printer filament, see "Printers, Filament, and Parts" on page 205.

Polylactic Acid (PLA)

PLA is available in many colors and can be opaque or translucent. A popular choice for 3D printing, it is plant-derived (corn or potatoes) and biodegradable. LayWoo-d3, Lay-Brick and FlexPLA are all specialized varieties of PLA. All varieties of PLA also adhere well to heated kapton or glass at 60°, which produces a smooth bottom surface finish on the printed part.

Nozzle temp:	185–235°C
Bed temp:	Ambient to 60°C
Print surface:	Blue painter's tape, heated glass, Kapton tape, sign cutting vinyl

Polylactic Acid (Soft/Flexible PLA)

Soft PLA is rubbery and flexible when printed, but comes in limited colors. For best results, print at a lower printing speed than regular PLA.

Nozzle temp:	210–240°C
Bed temp:	Ambient
Print surface:	Blue painter's tape, heated glass

LAYWOO-D3

This filament looks and smells "like wood" (made from 40% recycled wood and a binding polymer) and comes a variety of shades. Vary the print temperature for a cool effect: it's lighter at low temperatures, darker at higher ones. LayWoo-d3 may leave threads behind during non-extrusion moves of the print head.

Nozzle temp:	175-250°C
Bed temp:	Ambient
Print surface:	Blue painter's tape

LAYBRICK

This filament has a rough texture that looks similar to sandstone when printed. It can be brittle. Use a print temperature of 165-190 for a smooth finish and 210-230 for a rough finish.

Nozzle temp:	165-230°C
Bed temp:	Ambient
Print surface:	Blue painter's tape

Acrylonitrile Butadiene Styrene (ABS)

ABS is the plastic used in LEGO bricks and comes in a rainbow of colors. A commonly used 3D printing plastic, it requires a heated bed for proper adhesion.

Nozzle temp:	215–250°C
Bed temp:	90–115°C
Print surface:	Kapton tape

High Impact Polystyrene (HIPS)

HIPS can be used for printing final parts or as Limonene dissolvable support material. It prints better than and is much cheaper than PVA. HIPS prints have great surface finish to them that helps hide the print lines. For more on how to use HIPS as a support material, see Matt Stultz's post (*http://www.3dppvd.org/wp/2013/02/soluble-support-material/*).

Nozzle temp:	220-235°C
Bed temp:	115°C
Print surface:	Kapton tape

Nylon

Easily dyed (see Chapter 12), but can be difficult to use due to shrink/warp/curling problems. Good for both strong and low-friction parts and is flexible when printed in thin layers.

Nozzle temp:	235-260°C, but bonds best at 245° C
Bed temp:	Ambient
Print surface:	Scored nylon sheet, Garolite

Polyethylene Terephthalate (PET)

A crystal-clear, colorless filament that is strong and impact-resistant. Printing at thicker layer heights results in better optical clarity.

Nozzle temp:	210-220°C
Bed temp:	Ambient-65°C
Print surface:	Blue painter's tape, Kapton Tape, glass

Polycarbonate (PC)

Printing with polycarbonate requires high-temperature nozzle design, like the Prusa nozzle. This filament is considered experimental.

Nozzle temp:	280–305°C
Bed temp:	85°–95°C
Print surface:	Kapton tape

High-density Polyethylene (HDPE)

HDPE is difficult to use due to shrink/warp/curling problems and is rarely used.

Nozzle temp:	225–230℃
Bed temp:	Ambient
Print surface:	Polypropylene sheet

Polycaprolactone PCL

Also known as MakerBot Flexible Filament, PCL is a biodegradable polyester. It has a very low melting point (58-60° C) and can be heated in hot water and reformed. It is also commonly known as InstaMorph or Polymorph.

Nozzle temp:	100° C
Bed temp:	Ambient
Print surface:	Acrylic

Polyvinyl Alcohol (PVA)

PVA is sometimes used as a support material and dissolves in water. It is expensive and can be difficult to work with.

Nozzle temp:	180–200℃
Bed temp:	50℃
Print surface:	Blue painter's tape

There are some new types of PVA that will be available soon that will require different extrusion temperatures. The temperatures listed are for PVA that is currently on the market at the time of this writing.

Matt Stultz is the is the leader of the 3D Printing Providence group, founder of HackPittsburgh, and a MakerBot alumnus, with experience in multimaterial printing and advanced materials.

Industrial Materials and Methods

9

A materials guide for 3D printing services.

WRITTEN BY **STUART DEUTSCH**

There has never been a better time to purchase a desktop 3D printer. Nonetheless, they are still too expensive for many users. If you can't justify the cost of a personal printer, you may be able to access one at a local hackerspace, and there are many online 3D printing services to choose from, including Ponoko, Shapeways, and i.materialise. These companies use a variety of printing technologies to create physical objects from your digital designs and can print in many other materials besides extruded thermoplastic.

Composites and Ceramics

Powder bed and inkjet printers use inkjet-type print heads to deposit tiny droplets of liquid binder on top of a thin layer of powder. Once the build platform lowers, a roller spreads and compacts a fresh layer of powder across the surface. The final object is essentially a stack of powder layers finely glued together. Dyed binders can be used in certain machines to produce full-color display models (Figure 9-1). Treatment with super glue and UV protectants can improve model strength and reduce color fading.

Figure 9-1. *Bowie the Bunny in fine mineral power, with color binders, via power bed/inkjet process*

The powder bed/inkjet system can also be used to create food-safe ceramic models (Figure 9-2). The use of ceramic powder has become quite popular with online printers who now offer a rainbow of single-color options. After removal from the powder bed, raw ceramic parts undergo a series of heat treatments to dry, fire, and glaze the model, improving both strength and appearance.

Figure 9-2. *Food-safe ceramic, via powder bed/ inkjet process, followed by heat treatment*

Figure 9-3. *Fused nylon powder, via selective laser sintering*

Plastics

Stereolithography (SLA)

Stereolithography is the original 3D printing process, in which a liquid plastic resin is selectively hardened by exposure to high-intensity light, often from a laser. After the laser has drawn a 2D path along the surface, the freshly polymerized model layer is lowered into the surrounding resin bath. The laser traces over the fresh surface, curing and joining the resin to the previous layer. SLA produces prints of exceptional smoothness.

Selective Laser Sintering (SLS)

Selective laser sintering uses a high-power laser to melt and fuse particles of very fine plastic powder, often nylon (Figure 9-3). The laser scans across a leveled and compacted powder surface, and when each layer is completed, the entire bed is lowered and fresh powder is spread on top. As the laser works its way across the new layer, molten powder particles in the top surface fuse to each other and to the layer below. The unfused powder acts as support material, so SLS fabrication works well for models that have thin sections, overhangs, or complex geometries.

SLS is one of the most economical 3D printing methods and is forgiving in terms of design guidelines. Most vendors charge by volume of powder consumed, so you can often save money by "hollowing out" solid models and printing them as shells. Most vendors also charge for "trapped" powder, however, so models printed as shells will usually need to include at least one small hole so that the powder can be recovered when the print is done.

Photopolymer Jetting

Photopolymer jetting uses movable heads, like an inkjet printer, to deposit droplets of resin onto a build platform through a number of very small jets. Once the droplets are in position, a UV lamp moves across the platform to harden the resin. A support material may be printed surrounding the droplets and can be removed, manually or by washing, once the print is complete. Photopolymer jetting can create very finely detailed models with smooth surfaces and multiple materials—tinted, clear, rigid, flexible, etc.—in a single print (Figure 9-4). It is not widely available from 3D printing services, yet.

Figure 9-4. *UV-curing acrylic, via photopolymer jetting*

Metals

Direct Metal Laser Sintering (DMLS)

Direct metal laser sintering uses a laser to directly fuse certain metal powders, such as titanium, in a fashion very similar to SLS in plastics. Other specialty alloys can be printed via DMLS, but high costs and stricter design guidelines make the process less accessible to beginners.

Direct Metal Printing

Direct metal printing uses a multistep method to create powder-based metallic models, mainly from stainless steel. First, the object is printed into a bed of very fine stainless steel powder using the inkjet-binder process.

A carefully controlled heat treatment then burns out the plastic binder and fuses the steel particles together. Finally, the porous sintered model is infused with molten bronze, which wicks into the empty spaces and fills them (Figure 9-5). The finished model is a kind of stainless steel sponge filled with bronze, and can be given a variety of surface treatments, including plating with gold or other metals.

Figure 9-5. *Bronze-infused stainless steel, via direct metal printing*

Indirect Printing Methods

Indirect printing methods create positive or negative models that can be used with conventional casting processes to create metal parts. For instance, a sacrificial model of a part can be 3D printed in a wax-like resin using stereolithography, and then duplicated in metal using the traditional lost-wax process. Alternately, the powder bed/inkjet process can be used to print molds in silica sand or other traditional foundry media, which are then used to cast metal parts in the normal way.

Dr. Stuart Deutsch is a materials consultant in the NYC area and executive editor at ToolGuyd.com.

3D Printing Without a Printer

10

How and why to use 3D printing services instead of a desktop machine.

WRITTEN BY **COLLEEN JORDAN**

I'm incredibly lucky to have a job that even two years ago I never would have imagined could exist. I create 3D-printed jewelry and own a business, Wearable Planter, all thanks to tools and technology that have only been available for a few years.

When I studied industrial design at Georgia Tech from 2006–2010, 3D printing wasn't a tool we used frequently. We learned how to use 3D design programs to mock up our projects, but we typically used the resulting 3D files for product rendering. Of course we had access to a 3D printer, but few people knew how to use it, and the models it made were fragile and expensive.

It wasn't until my last semester of college that I worked on a project that required 3D printing. I handed the flash drive with my files to the lab assistant, thinking it would never work, and then watched with amazement as my jewelry piece was printed layer by layer exactly as I had intended.

Most people with the desire to imagine and prototype a new product don't have immediate access to a professional-quality 3D printer. The good news is that access to this technology is increasing at an unprecedented rate.

When I design a new product or piece of jewelry, I begin by creating sketches of what it will look like. This part of the process often takes the longest, as I am deciding on the form and feel of the object. I then create a 3D model in SolidWorks, Rhino, or whichever program I feel will allow me the most creativity with my knowledge of the software. After I'm done creating, I export the file for printing, and I run checks with a program like Netfabb Studio to make sure it's suitable for printing.

I then use 3D printing services like Shapeways and Ponoko—both to prototype and to create my final products.

When I use these services, I have to wait two weeks or longer to see how my print turned out. That can seem like a long time, but it affords you a new perspective when you're forced to step away from your project momentarily. Sometimes when I get my first print back it's exactly what I was expecting, but often I see changes that need to be made, whether in material choice or wall thicknesses or other details.

The fantastic thing about using 3D printing services is that I'm able to run a business with tools that these companies have created—to experiment with different materials and new products inexpensively and with little overhead. In the past, creating a prototype could be costly, and production of even a limited run of items could be thousands of dollars. Also, I'm able to keep a limited inventory on hand because I can restock it quickly. This beats previous business models where I might have needed to order a cargo container full of pieces from overseas.

There are more advantages to using 3D printing services than just running a business. We're coming into a new era where mass customization is driving the production of new objects. Access to these technologies is now open to everyone, from beginners making character models with Minecraft to doctors making personalized prosthetics.

For example, if you'd like to create your own phone case or dishware, but don't have any 3D modeling experience, you can use one of the many "creator" programs to make your own customized item. Sculpteo and the Society for Printable Geography recently released an app to create iPhone cases with the terrain of your favorite place. Shapeways lets you create a sake set by playing around with the shape of curves. These apps let you create your own unique manufactured item with minimal cost.

Portable Greenery 3D-Printed Mini Pots from Wearable Planter

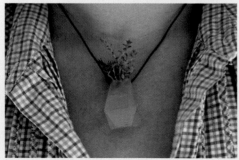

These companies have also begun to offer training so you can learn to use their tools. Ponoko offers online training classes, aimed at beginners, that are free to watch and participate in. Shapeways offers Skillshare classes to teach introductory skills as well as more complex, generative software.

Sending your work off to 3D printing services has many advantages over using equally popular desktop 3D printers. You won't have the up front investment of $300 to $2,000, and you won't spend time tinkering with settings and hardware. You also can create more complex and higher-resolution objects, as desktop 3D printing still doesn't match the quality of professional machines.

The technology does come with its disadvantages. As this is a rapidly growing field, sometimes the demand is greater than the capabilities of these companies, and unexpected delays in lead times arise. The quality of materials isn't always as good as a similar mass-produced piece; 3D-printed plastic parts, for example, can be more fragile than a similar injection-molded piece. And, importantly, not all of these materials are food-safe or suitable for use in toys. While these disadvantages may discourage you, keep in mind that this amazing technology wasn't available to the public even five years ago, and it's developing very quickly.

If you have an idea you'd like to bring to life, there's no better time to see what may come from it. Check out "3D Products Now on the Market" on page 153 to see see how all kinds of makers are taking advantage of 3D printing services, Chapter 11 for where to have them made, and Chapter 9 for a rundown of available materials.

Colleen Jordan is a designer and maker who likes to create objects that make life more interesting. She is the founder of Wearable Planter (http://wearableplanter.com:), and dreams of one day having a pet dinosaur.

Service Providers

11

These companies will print your models in a range of exotic materials.

COMPILED BY **COLLEEN JORDAN, STETT HOLBROOK,**

AND ANNA KAZIUNAS FRANCE

If you don't have a 3D printer, you can still enjoy the benefits and fun of turning a CAD file into a physical object. It's well worth getting familiar with the growing number of 3D printing service providers, and commercial providers can print in a surprising range of materials not available to the home printer user. Even if you do own a printer, it's great to be able to turn a piece of art or technology you prototyped in plastic into something enduring, like titanium or stainless steel.

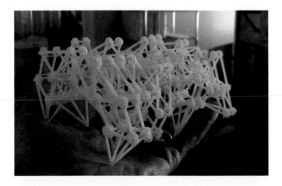

Figure 11-1. *A Strandbeest mechanism by Theo Jansen, printed in nylon on an EOS selective laser sintering (SLS) machine at Shapeways headquarters in New York. Dust it off and it's ready to walk.*

Upload Files and Order Prints

Shapeways

http://shapeways.com

Shapeways caters to hobbyists and designers, offering high-quality prints in a wide range of materials, including sterling silver, stainless steel, brass, and ceramic. They often have the lowest prices, but with production based in the Netherlands, their stated shipping times of 2–3 weeks haven't always been accurate. They recently built a production facility in New York to meet demand.

Shapeways also offers a marketplace of designs where users can open their own shops. Because theirs is the most visible of these services, having a shop on their site is almost essential to getting started selling your work. There's little up-front cost or commitment.

Ponoko

http://ponoko.com

Ponoko offers 3D printing in a wide range of materials, from plastics and ceramics to stainless steel, gold plate, and Z Corp plasters. They also offer laser cutting and CNC routing in a huge variety of materials, so you can supplement your 3D-printed project with other custom parts.

Prints from Ponoko are generally very good quality and reasonably priced, but their pricing structure and system for uploading models are confusing. They do have a very good support staff who will go above and beyond to help you with any questions. Ponoko operates several regional production facilities, so printing and shipping times vary.

Sculpteo

http://sculpteo.com

Material choices from this France-based company include multicolored plastic, resin, ceramic, wax, alumide, and sterling silver. The company sells a sample kit of materials for $5. They offer a preview and live printability check that is very useful, and automatic model repair inside the web interface. They also offer many "creator" apps for you to begin making right away. Shipping times vary from 1–30 days depending on the material (upload your files for real-time estimates). Users may also post their objects for sale on the company's website.

i.materialise

http://i.materialise.com

Another 3D printing service based out of the Netherlands, i.materialise has a very clear and easy-to-use interface. Choose from more than 20 materials—including titanium!—and print objects as large as 6 feet.

Objects you create can be sold on the company's gallery. Depending on your location, objects ship in 1–5 days.

Kraftwurx

http://www.kraftwurx.com

Kraftwurx is a platform that enables individuals to create, buy, sell, and display 3D-printed products. Headquarted in Houston, TX, Kraftwurx does not own a factory but instead uses a network of over 120 manufacturers for local on-demand production. This distributed production model allows them to provide 85 different materials, including gold, sterling silver, titanium, Iconel, stainless steel, Platinum, plastics, and paper.

Staples (partnering with Mcor)

http://staples.myeasy3d.com/

Myeasy3D is a collaboration between Staples and Mcor, the producer of the Iris full color paper 3D printer. This service is currently only available in Europe.

Makers Producing Parts Locally

makexyz

http://www.makexyz.com

makexyz connects local 3D printer owners and people who need 3D prints. If you have a printer, registration is free and you can also advertise design services. If you need prints produced, you choose a listed printer near you, upload your 3D file, configure your options, and then check out through makexyz. The printer can either accept or decline your order, and you can have your print shipped or pick it up.

3D Hubs

http://www.3dhubs.com

3D Hubs works in a similar way to makexyz: it aims to facilitate transactions between printer owners with those who want prints, but is also focused on building community and critical mass as "local print hubs" in an area before it begins connecting makers. Check out the web site for available 3D Hubs near you.

Find Me a Printer

Print Chomp

https://www.printchomp.com

Print Chomp provides printing quotes for a variety of printed goods, from 3D printing to business cards to banners. You upload your file and configure your options and they send you a quote.

Professional-Grade Services

ZoomRP.com

http://zoomrp.com

Here's a self-service website for those who need plastic prototypes in a hurry. The site, a division of Solid Concepts, specializes in same-day 3D printing and shipping, with next-day and next-week options too. It's best suited for customers who know their way around CAD and 3D design. Fast, but it costs more.

RedEye

http://redeyeondemand.com

RedEye, a division of 3D printing giant Stratasys, serves the architectural, medical, engineering, and aerospace industries with rapid prototyping. The company prints objects in various thermoplastics, UV photopolymers, and resins. Professional services, professional prices. Shipping time ranges from 1–2 weeks.

3D Factory

http://www.3dfactoryusa.com

3D Factory is based in NYC and focuses on the jewelry industry. Their services include 3D printing, casting, stone setting, finishing, and 3D scanning services. 3DPhactory also provides consulting and design services.

Boutique 3DP Design and Printing

3dPhacktory

http://3dphacktory.com

3dPhacktory is located in Toronto, Canada, and provides both 3D printing and design services that can ship printed designs within 48 hours. They offer 24-hour rapid prototyping turnaround, personalization, and 3D scanning services through Industrial Pixel. Customers are encouraged to visit the 3DPhacktory collaborative with their designers or rent time on a workstation and design themselves. They also provide post-production services.

Solid-Ideas

http://www.solid-ideas.com

Solid-Ideas is a boutique model-making and design agency in Northern California that offers model-making, prototyping, and custom fabrication services. They specialize in professional services for architecture, marketing, and other fields that take products from conceptual design to prototypes and marketing models.

Colleen Jordan is a the founder of Wearable Planter (http://wearableplanter.com).

Stett Holbrook is a senior editor at MAKE.

Anna Kaziunas France is the Digital Fabrication Editor at Maker Media.

PART VI
Finishing Techniques

How to Dye Your 3D Prints | 12

How to add color to your nylon (or polyamide) prints with nylon with fabric dye.

WRITTEN BY **COLLEEN JORDAN**

Have you created something with 3D printing? Many design students and hobbyists now have access to the technology thanks to services like Shapeways and Ponoko. If you print your objects in polyamide, you can dye them at home to whatever color you want. Polyamide is a porous material that accepts color really well. Some companies offer dyeing of your prints for you, but that adds extra processing time and is only available in a small range of colors.

If you're tired of the boring white that many 3D prints come in, we will show you how to add color to your prints (Figure 12-1). This is a tutorial for dyeing nylon (or polyamide) 3D prints with fabric dye. This material is known by different names at different printing companies. Shapeways calls it "White Strong and Flexible", Ponoko calls it "Durable Plastic", Sculpteo "White Plastic", and iMaterialise "Polyamide". We'll use Rit brand dyes in our tutorial since they are easy to find in craft, fabric, and grocery stores. You can also dye your prints with Jacquard brand acid dyes with a similar process, but that will require carefully measuring vinegar to change the acidity of the solution and constantly heating the solution.

Figure 12-1. *The author's Wearable Planter*

This process is similar to dyeing fabric, and we learned a lot about how to dye 3D prints by reading this article on dyeing techniques by Rit (*http://www.ritdye.com/dyeing-techniques/microwave*).

If you have a desktop 3D printer, you can dye filament (see "Nylon" on page 96) prior to printing to achieve a "tie-dyed" look. Check out RichRap's seminal tutorial (http://richrap.blogspot.co.uk/2013/04/3d-printing-with-

115

nylon-618-filament-in.html) on *how to dye nylon filament.*

1. Gather Your Materials

The first thing that you will need to do is gather your materials (Figure 12-2). You'll need your nylon 3D prints, your desired color of fabric dye, a bowl to do the dyeing in, measuring spoons, and boiling water (not pictured). We also recommend having access to a microwave to reheat your solution while dyeing as needed.

Figure 12-2. *Supplies*

Decide which color you would like to dye your prints. Rit has a great guide to tell you which colors you can dye your prints with (*http://www.ritdye.com/colorit_color_formu la_guide*); other brands of dye will have similar guides. Nylon absorbs the dye really quickly, and we usually use slightly less dye than the guides recommend. For this batch of bike planters we will be dyeing them using Rit's Sunshine Orange. We're using 1.5 tsp of powdered dye to 1.5 cups of boiling water.

Remember that you are working with fabric dye that will stain clothes and shoes. So if you care about the clothes that you are wearing,

wear an apron or change into something that you don't love so much. Fabric dye can also stain your skin, so wear latex gloves if you don't want tinted hands. Rit dye will come off easily with scrubbing, so if you do get some on your skin, it can be easily removed.

2. Soak Your Pieces

Before you begin the dyeing process, soak your prints (Figure 12-3) for at least 30 minutes. We recommend doing this overnight if you have the time. Having your prints saturated will allow the dye to color the pieces more evenly. This will also help remove any dust on the surface of your prints left over from the printing process. If there is residual powder on the surface of your prints, it will affect the color of piece. The powder will be dyed and will come off easily when the piece is dry, leaving a white spot underneath.

Figure 12-3. *Soaking*

The piece shown in Figure 12-4 had some leftover powder stuck to it when it was dyed, and you can see the large white area left behind from removing the powder.

Figure 12-4. *A blotchy piece*

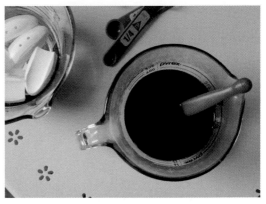

Figure 12-6. *Stirring it well*

3. Add Color

Carefully measure your required amount of dye (Figure 12-5) and add your boiling water. Stir it really well so all of the powder is dissolved in solution (Figure 12-6).

Figure 12-5. *The dye in the container*

Add your prints to the solution and stir (Figure 12-7). Agitate the solution frequently to ensure that your prints are colored evenly. The longer that you leave your prints in the solution, the more saturated the color will be. These prints stayed in the dye for about six minutes to achieve the color shown. If you need to leave your prints in the solution longer, microwave it at 15- to 30-second increments to reheat the water to near boiling temperature. We've noticed that some dyes require higher temperatures to stay in solution than others. In our experience pink and blue dyes require hotter temperatures and longer dyeing times to achieve their desired colors.

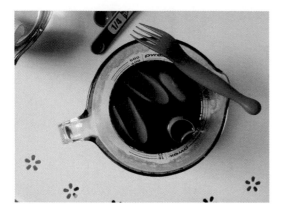

Figure 12-7. *Adding the prints to the solution*

4. Rinse

Rinsing your prints is very important (Figure 12-8). You can rinse them with cold water to remove the excess dye. We also like to let the pieces sit in boiling water for a few minutes for any excess dye to soak out. If you're going to be dyeing jewelry or anything that will be worn close to the skin, this is a very important step as excess dye could stain the skin or clothes.

Figure 12-8. *Rinsing*

5. Dry

Next, dry everything out (Figure 12-9).

Figure 12-9. *Drying your prints*

Optional Step: Seal

Nylon is a porous material that will readily absorb particles and dirt it is exposed to. We recommend sealing your prints with a polymer varnish (like Liquitex) or clear acrylic paint to protect the color and your piece from getting dirty.

6. Show It Off!

You just put all this hard work into your 3D-printed object—show it off (Figure 12-10) and tell everyone about it!

Figure 12-10. *One of the bike planters in action*

This post was originally published on the Wearable Planter blog (http://www.wearableplanter.com/blogs/news).

Colleen Jordan is a designer and maker who likes to create objects that make life more interesting. She is the founder of *Wearable Planter* (http://wearableplanter.com), and dreams of one day having a pet dinosaur.

Post-Processing Your Prints

13

Friction weld, rivet, sand, paint—arm yourself with simple tools and finishing techniques to take your 3D prints to the next level.

WRITTEN BY **MATTHEW GRIFFIN**

PHOTOGRAPHY BY **ANDREW BAKER**

People often claim 3D printers can "make you anything you can imagine." Dial up the digital model you want, hit "Go," and the machine hums to work, producing an object accurately and repeatably. But as an astute eight-year-old pointed out to me when I handed her two of my favorite printed models at Maker Faire Bay Area last year, the results don't always match your intentions.

"That octopus is *red*! A TARDIS is not *supposed* to be yellow!" she wailed, and knocked my offerings away.

While overall shape and mechanical fit are valued more highly than surface treatment in today's desktop 3D printing, it's sometimes worth judging a print by its cover.

I'm reminded of advice I got from a pair of industrial design professors at Pratt, after I showed them my print of a fluorescent-green clockwork mechanism: "It is worth enormous effort to make prototypes look like they were created from real-world materials." Even the most creative engineers

and business people will have difficulty seeing your prototype as a *machine* when it looks like a toy.

The domain of finishing techniques (i.e., everything that takes place after printing) is the craftsman's workshop, where patience, tools, skills, and experience can transform the raw products of these machines into fully realized models. Like builders of dollhouses and model trains, many 3D printer jocks appreciate a loving and accurate rendering of a miniature world.

The results are impressive, but why should you tackle these craft skills when you could spend that time printing more plastic objects?

Makers who have mastered finishing techniques are granted wizard status by fellow 3D practitioners. Take artist Cosmo Wenman, who creates pieces that accurately mimic distressed metals and stones, and sculptor Jason Bakutis, whose sanded, painted, and polished faux marble and jade prints look

121

remarkably like the real thing. Through careful work, pieces printed in crazy pink, green, and translucent filaments are made to resemble clay, stone, metal, and wood. How do they do that?

Tools and Materials

- Benchtop vise such as a PanaVise
- Pliers, combination (aka lineman's pliers)
- Pliers, needlenose
- Multitool
- Safety goggles (I like DeWalt's DPG82-11C clear anti-fog model)
- Respirator for sanding/particulates (I use 3M's 8511 particulate respirator)

For friction welding:

- High-speed rotary tool with 1" and 3/32" collets, such as a Dremel
- Filament for 3D printer, ABS or PLA

For heating/reworking:

- Hot air SMD rework station (Figure 13-1) or other small heat gun
- Soldering iron and solder
- Brass tube to fit snugly over your soldering iron tip
- Metal plate or mirror (optional) for fast cooling
- Nail, steel, large (for cooling/pressing)

For trimming/grinding:

- Deburring tool (I use Noga's heavy duty NG-1 model)
- Flush cutters (aka diagonal pliers) or wire cutters

- Files, narrow, diamond grit
- Coffee/spice grinder (for grinding filament)

For sanding/polishing:

- Sandpaper: 80/100, 150, 220, 320, and 500 grits
- 3M Wetordry Polishing Papers
- Micro-Mesh Soft Touch Pads and Colored Sanding Sticks
- Sanding/polishing/buffing disks for rotary tool
- Novus Plastic Polish Kit

For filling/gluing/painting/sealing:

- Acetone for use with ABS objects (not PLA)
- Resealable container, acetone-resistant
- Enamel hobby paints such as Testors
- Acrylic paints
- Clear-coat spray paints (I use Krylon Crystal Clear Acrylic, Matte Finish, and Triple Thick Crystal Clear Glaze; and Rust-Oleum Matte Clear and Gloss Clear).

Figure 13-1. *Hot air rework station*

The desktop 3D printing community has a lot to learn from the sculptors, model railroad builders, and tabletop gamers now joining their ranks. And as my professors pointed out, these extra steps aren't just cosmetic. Your capacity to transform your models into "magical" replicas is a crucial means of communicating your inventions.

Tricks of the Trade

Desktop 3D printing has yet to spawn third-party finishing services like commercial 3D printing did a decade ago. So, without access to acetone cloud chambers, multiaxis enamel jet robots, agitating chemical baths, and industrial tumblers and polishers, makers have rolled up their sleeves and discovered a host of finishing solutions using inexpensive tools and materials. These methods not only affect a print in post-production, but can often change the way we think about a digital model back in the initial design stages.

In researching my upcoming book *Design for 3D Printing* (MAKE, 2013), I've interviewed a wide range of members of the desktop 3D printer community. I'd like to share some of their promising tools and techniques. In turn, I hope that those of you refining new methods and sourcing better, safer, and cheaper products and techniques will also share.

Friction Welding

The world may have forgotten the Spin Welder toy sold by Mattel in the mid-1970s, but Fran Blanche of Frantone Electronics did a great job of recreating the experience in her 2012 video "Build Your Own Friction Welder." Using an inexpensive rotary tool, Fran was able to spin a styrene rod fast enough to create a strong weld between two pieces of plastic that was difficult to break apart by hand. With the Spin Welder toy, children assembled the frames of helicopters, motorcycles, and other projects by fusing together beams and struts, then used plastic rivets to fasten the outer shell. Sure, it was potentially one of the most dangerous toys of all time, but I agree with Fran's conclusion: why haven't tools like these joined the maker's toolbox?

Unlike adhesives or traditional welding, friction welding fuses metal or thermoplastic objects together by quickly spinning or vibrating one piece against another. Mechanical friction creates a melt zone shared by both parts, fusing them into one solid piece. In friction surfacing—a variant of friction welding—a piece rotated at high speed is moved across an edge or surface under gentle pressure to weld seams, patch gaps, or smooth surfaces.

These techniques are common for plastics and aluminum in the automotive and aerospace industries, but the tools are expensive. Sophisticated spin welders can spin parts at hundreds of thousands of RPMs for short bursts of even single-digit rotations, parking the fused part at a precise orientation. Where are the cheap, hand-tool equivalents?

As it turns out, many of us already have the equipment to experiment with friction welding. Dremels and similar high-speed rotary tools spin fast enough to melt 3D printer plastics, and printer filament can be used as welding "rod" to solidly fuse parts or close seams. These tools can also spin-weld 3D-printed rivets. And while it takes them a second or two to spin down again, the melting points are comparatively low, allowing for some manipulation after the fact to reposition the joined part.

I think both approaches—welding and riveting—are killer tools for 3D print finishing, particularly for "blind riveting" into the side of an object, and for joining parts made of PLA, which is typically much harder to glue than ABS.

Friction Welding Mismatched Surfaces

Figure 13-2. *Preparing the rotary tool*

I spent some time with Chris Hackett from the Madagascar Institute learning how ideas from traditional metal welding might apply to friction welding 3D-printed parts. We experimented with the rotary tools in his workshop and came up with the following approach for creating a nice welded seam in plastic, similar to a traditional metal weld. When two printed parts don't mate perfectly due to warp or poor planning, you can friction-weld them together as securely as if they were a single printed part. Here I'll demonstrate with ABS parts and ABS filament. It works with PLA, too.

1. Prepare the Rotary Tool

Select the collet you need for trapping the filament you'll be using. For 1.8 mm filament, use a 3/32" collet as shown here (three rings) and for 3 mm filament use a 1" collet (0 rings) (Figure 13-2).

Insert a short length of filament into the collet jaws and tighten down the collet nut to secure it in place (Figure 13-3).

Figure 13-3. *Insert filament into tool*

Trim the filament about 1/2" from the collet. Short pieces are easier to control, and they spin on a tighter axis. (With experience you can use longer pieces, pressed gently at an angle, to make longer welds. You may need to straighten them by reforming them with a heat gun.)

2. Prepare Two Parts for Welding

After scraping and sanding, the two watch body cases shown in (Figure 13-4) meet with a gap that varies between 0.1 mm and 2 mm around the edges. That's too sloppy for gluing, so I'll weld them.

Figure 13-4. *Prepare parts for welding*

Use a deburring tool or razor blade to bevel the top edges of the seam where the parts meet, forming a narrow, V-shaped channel (Figure 13-5). Your goal is to create enough room for three welding layers, from the bottom of the bevel up to just above the surface of the two parts. This method gives a stronger bond than a weld that sits just on the surface.

Figure 13-5. *Bevel the top edges of the seam, forming a narrow, V-shaped channel*

Warm both parts with low heat from a heat gun (Figure 13-6). This helps them receive the weld to the same depth. If one part is much larger than the other, focus extra attention on warming the larger piece.

Figure 13-6. *Warm the parts with a heat gun*

3. Tack-Weld the Parts in Position

Now tack the parts together with a series of short spot welds, moving around the joint while holding the parts steady.

Spin up the rotary tool, and lower it until the spinning filament grazes both surfaces of the seam. When the tip of the filament begins to deform, apply a little pressure (Figure 13-7).

Figure 13-7. *Applying pressure*

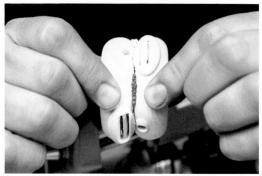

Figure 13-9. *Tack the seam and let it cool*

Moving the spinning filament in tight little circles, widen the melt zone slightly into the sides of both parts, making a little forward progress with each circuit, until you've created a small spot weld (Figure 13-8).

4. Plug Gaps with Filament

Gaps that are wider than half the width of your welding filament should be filled before welding a clean seam. Soften a short scrap of filament to use as filler, by using the low setting on a heat gun or by warming it to 100°C on your printer's heated build platform (Figure 13-10).

Press the softened filament into the widest gaps between the two parts, making quick tack-welds if necessary to pin it in place.

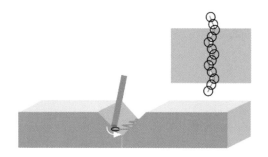

Figure 13-8. *Spot welding diagram*

Tack in three or more places along the seam and let the parts cool. They should be tacked tight, difficult to separate by hand (see Figure 13-9).

Figure 13-10. *For large gaps, soften a short scrap of filament to use as filler*

5. Friction Weld the Seam

Now weld the whole seam in two or three layers, as shown in Figure 13-11. A single weld would probably bond the parts at the surface only, allowing the seam to be broken if the parts are torqued.

In this idealized diagram, we weld one bead at the bottom of the seam, two beads on a second layer, and three on a third (top) layer, fusing the parts through their entire thickness.

To hide the weld, you can sand it back flush and then paint or seal the surface.

Friction Welding to Repair a PLA Model

Punctured mysteriously by the TSA (Figure 13-12), a part from Micah Ganske's sculpture Industrial Ring Habitat (see Figure 13-13 for context) needs a patch. We'll use red filament to make the friction weld easily visible.

Figure 13-12. *Punctured "Ring Habitat"*

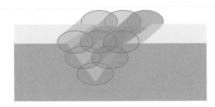

Figure 13-11. *Friction welding the seams, using several layers*

Figure 13-13. *Industrial Ring Habitat, by Micah Ganske*

PLA is prone to cracking and splitting, and it's typically difficult to repair. Solvents such as acetone have little effect. ABS glue and super glue merely cement the parts by surface tension, rather than offering a chemical weld—meaning that the seam can easily be rebroken.

With friction welding, you can form solid joints that are difficult to break.

1. Press the patch or broken part into place and hold it securely:

2. Tack-weld the patch in place in a few different positions:

3. PLA is workable at lower temperatures than ABS, so use a gentle touch to melt and weld the two parts. Too much pressure can create a puncture. It takes a bit of practice:

4. Move around the seam, changing direction as necessary for handling and control. For prettier welds, take frequent breaks to let the parts cool:

5. Let the completed seam cool to room temperature before sanding and sealing:

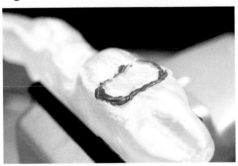

Riveting: Friction Welding Blind Rivets

A rotary tool can also be used to permanently fuse a spinning part to a fixed one, using a one-sided "blind rivet." Blind rivets have one huge advantage over ordinary solid rivets: you don't need access to both sides of an assembly to rivet its parts in place.

This method works well for attaching plastic panels to the outside of objects when access to the interior is awkward or impossible. It also lets you construct massive objects from multiple panels, each panel printed close to the bed of the printer for optimal printing.

Shown in Figure 13-14 are two 3D-printed blind rivets, next to a brass solid rivet and three aluminum blind rivets. Notice that the printed rivets, like the aluminum ones, have a "mandrel" that extends well beyond the rivet's head. This is the part that's gripped in the rotary tool.

Figure 13-14. *3D-printed blind rivets, brass solid rivet, aluminum rivets*

You'll clip it off after the rivet is firmly in place. I designed the printed rivets to be gripped by the 1" collet, the standard size for most Dremel accessory bits.

Plastic rivets need not be perfectly cylindrical for friction welding, so I designed them "three-quarters" round, for printing flat on the platform (Figure 13-15). This way, the horizontal "grain" of the printed rivet helps strengthen it.

Figure 13-15. *Three-quarters round rivets*

I'll demonstrate by riveting a small panel to the outside of another part. To print your own rivets, get the 3D files at *http://www.thingiverse.com/thing:61510*.

 Friction welding involves the use of high-speed rotating tools and should not be attempted without ANSI-approved safety glasses. Welding and other operations that heat, soften, and melt plastic may release hazardous chemical vapors and should not be attempted without proper ventilation. Sanding and other dust-producing operations should not be attempted without a NIOSH N95-approved particulate respirator. Acetone and other volatile solvents should not be handled without proper ventilation, safety goggles, protective clothing, and latex or nitrile gloves.

1. With the proper collet in place, loosen the collet nut and insert a plastic rivet:

2. Drill or design in mounting holes in your panel to provide clearance for the shaft of the rivet to pass through to the base part where it will be fixed. The hole should be narrow enough that the rivet's head will pin the panel in place.

Pre-drilling (or designing) a pilot hole in the base part can help prevent your rivet from mounting off target:

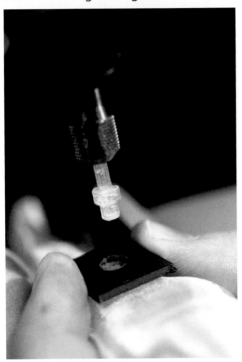

3. Spin up the rotary tool and gently insert the shaft of the rivet through the mounting hole until it contacts the mounting position. Continue spinning until the shaft of the rivet begins to melt and deform—then press it gently down into place:

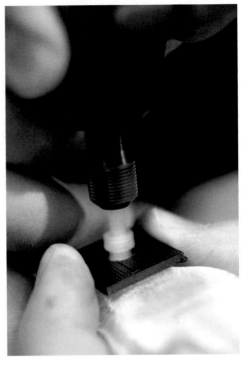

4. Stop the rotary tool and hold it steady in a fixed position at a right angle to the work, while applying a little downward pressure. It can help to use a piece of cardboard or foam as a friction brake to stop the rotation quickly (unlike a professional spin-welding tool, most rotary tools need a second or two to spin down).

5. Loosen the collet nut and slip the mandrel of the blind rivet out of the rotary tool. If the rivet is still cooling, hold it in position until it's fully cooled (at which point it should be entirely fused with its mounting point):

6. Using a flush cutter, snip off the mandrel, leaving the head intact:

7. If the rivet head protrudes too far, has a sharp ridge, or seems too narrow to secure the panel in place, warm it with a heat gun and use the head of a steel nail to press it flat:

It's possible to fuse ABS rivets to PLA, and vice versa, but you'll need to find the "feel" for the initial friction stage before pressing down the body of the rivet. Before mounting delicate parts, test-rivet the materials you'll be using.

Using Filament to Make Solid Rivets and Hinges

People have used rivets since the Bronze Age to fasten together tools, art, bridges, and buildings, so it's no surprise that 3D printer

users are experimenting with riveting techniques. We've seen a number of projects using pieces of filament as pins to hold together large assemblies.

Just recently, 3D artist and instructor Jason Welsh demonstrated a method for building his DIY electronics cases that promises to become a new power technique. His Folding Arduino Lab (*http://www.thingiverse.com/thing:32839*) (Figure 13-16) and Pi Command Center (*http://www.thingiverse.com/thing:38965*) each use filament "spikes" to create rivets and hinges.

Figure 13-16. *Pi Command Center*

Essentially, Welsh uses heat to reform pieces of filament into straight rivets, flattening one head before inserting the rivet and the other head after the rivet is firmly in place. As with any solid rivet, you need access to both sides of the assembly, but the advantage of this method is the creation of strong fastenings that can be completely removed later using a flush cutter.

While you can make spikes with any filament, I recommend 3 mm PLA based on my experiences building Welsh's project. PLA is easier to soften and work with a heat gun, and 3 mm spikes remain straighter and more rigid than 1.75 mm spikes after cooling. If you don't have 3 mm filament, you can accomplish the same goal with 1.75 mm filament by using more rivets to distribute the load.

1. With a heat gun set to low, evenly warm a 4"–6" length of filament until it becomes limp (several minutes on a heated build platform works, too):

2. While the filament is still hot, straighten it by rolling it on a table, or better yet, on a piece of glass that will quickly cool it. Gently move both hands away from each other while rolling, to keep the filament straight as it cools:

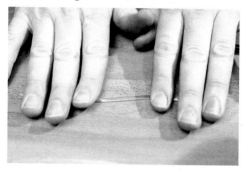

3. Soften one end using a heat gun on a low setting (a soldering iron, heated brass nozzle, or heated build platform can work in a pinch):

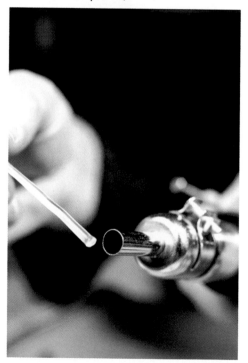

4. Stop the rotary tool and hold it steady in a fixed position at a right angle to the work, while applying a little downward pressure.

   It can help to use a piece of cardboard or foam as a friction brake to stop the rotation quickly. (Unlike a professional spin-welding tool, most rotary tools need a second or two to spin down.)

5. Tap the soft end of the spike on a flat, cool surface until it deforms into a flat rivet head. I tend to use a steel nail head, but any flat surface that can cool the filament rapidly will work:

6. Your rivet should have a nice flat head, wide enough to rest firmly on the edge of the mounting hole. In rivet lingo, this is the "factory head," as opposed to the second head or "shop head" you'll create on the other end when installing the rivet:

7. Insert the rivet into the mounting point until the factory head is flush, then clip the tail a little ways beyond where you need the second head of the rivet:

8. Use a heat gun to soften the protruding tail of the rivet until it begins to deform:

9. Use a flat, smooth surface to press down and deform ("buck") the tail, creating the rivet's shop head. I find that a large steel nail head works best—it's easy to handle and it cools the shop head quickly:

10. Continue to press tight, against the shop head as it cools, making sure it doesn't relax away from the mounting point.

Don't apply force to the assembly until both the rivet and the parts being assembled cool to room temperature.

Hold in place until cool as seen here:

1. Installation of a hinge rivet follows the same procedure, except at the very end.

After pressing the shop head, move the parts gently as they cool to ensure that the joint has enough play for the hinge to open and close easily.

In the completed assembly, the two black ABS plates can spin easily on the hinge without coming free.

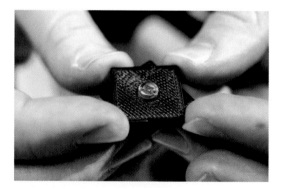

Two Materials, Two Approaches to Finishing Tech

ABS and PLA plastics have very different physical properties. ABS is printed at a higher temperature (typically 215–235°C), is more durable and flexible, and dissolves in industrial solvents like acetone. PLA can print at lower temperatures (starting at 180°C), wears down faster, can be brittle or shatter, and won't dissolve in acetone. (The chemicals used to dissolve PLA are highly toxic.)

If you'd rather use a soldering iron than a heat gun, find a brass tube (Figure 13-17) that fits snugly over your iron. Use the brass tube to work the plastic, and keep your soldering tip clean. Make sure to clean your tube thoroughly so the plastic doesn't stick to the brass.

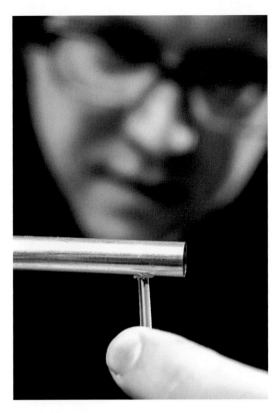

Figure 13-17. *Protect your tip with a brass tube*

Tape the factory head of the rivet in place when you need both hands to soften and flatten the shop head.

Gluing and Filling: Creating ABS Slurry for Filler and Glue

While super glue (cyanoacrylate) and plastic model glues do an excellent job of bonding ABS parts, many 3D-printed model builders have switched to using "ABS slurry" for both glue and filler material, because this substance can weld parts together more permanently and can be exactly color-matched to the printed parts. ABS slurry is simply ground-up ABS filament dissolved in acetone.

Applied in the open air, acetone melts the surface of ABS plastic (and many similar styrene plastics), creates a goopy sludge, and then—after some time—evaporates, leaving behind just the reformed ABS plastic. By sealing up this process in an airtight container that the acetone cannot easily escape, you can prepare a thick, even acetone/ABS mix similar to acrylic gel medium.

There are a variety of methods for preparing ABS slurry. I like ProtoParadigm's recipe (*http://www.thingiverse.com/thing:14490*); one part ABS to two parts acetone, mixed in fingernail polish containers or similar. Use a cheap coffee/spice grinder to shred ABS filament and scraps as needed. Smaller pieces dissolve faster and make it easier to gauge the mix ratio.

 Observe proper handling precautions when working with acetone and ABS slurry. Wear gloves and goggles and do not work without proper ventilation or in the presence of open flames. Besides being highly flammable, ABS slurry sticks to anything and burns with a foul-smelling smoke that is widely regarded as toxic. Be very careful or you'll create a tiny batch of "napalm" that will need to be treated like a chemical fire.

Apply ABS slurry with an inexpensive natural-hair paintbrush (synthetic brushes will dissolve in acetone!) to either fill small cracks or glue two pieces together. Leave it to air-dry until the acetone completely evaporates, and your final part will have a joint or patch made only of ABS plastic.

Your exposure to acetone is greatest while applying ABS glue and immediately after, so pin your parts to a piece of cardboard or a tray that you can immediately move to a well-ventilated area away from your workspace. If you move them outside, protect them with a cardboard box to keep leaves, dust, and grime out of the still-goopy slurry.

While acetone can "weld" the edges of ABS parts to bond them, this joint lacks the shear resistance of parts printed together, because the "melt zone" doesn't extend deep into the surface. If an assembly needs mechanical strength, design an interlocking joint with lots of surface area—or use hardware.

Sanding 3D-Printed Plastic Parts

When I first learned the basics of woodworking, I proved a lazy, inept sander of splintery plywood toolboxes and lopsided Pinewood Derby cars. My father suggested I forget about "sanded" as a goal, and focus on "sanding" as an activity. You cycle from coarse-grit sizes down to finer-grit papers until the surface is as smooth as you intend.

The same goes for sanding 3D-printed plastics (Figure 13-18). With ABS and PLA, you can work your way down to very fine papers indeed—3M gem-polishing papers and Micro-Mesh sanding tools with single-digit micron grits that create scoring patterns invisible to the naked eye.

Industrial Ring Habitat

Artist Micah Ganske used ABS slurry glue to assemble his groundbreaking sculpture Industrial Ring Habitat from 1,000 3D-printed ABS parts. He also uses it to glue PLA parts to ABS —even though acetone doesn't dissolve PLA, the slurry seeps into cracks and crevices to mechanically bond the PLA parts to the ABS base.

Figure 13-18. *Sanding 3D prints*

Still, well-sanded 3D-printed projects seem few and far between, for two good reasons. First, ABS and PLA are softer than the wood we're used to sanding. Second, the tricky horizontal "grain" created by 3D printing reflects light differently than sanded surfaces or the glossy heated base (in ABS printing), and this grain cannot be tooled back into the surface easily—encouraging an all-or-nothing approach to sanding the object.

The basic rule of thumb is to sand 3D-printed pieces like you'd sand a gummy hardwood. Focus on "sanding" and don't rush toward "sanded": start with 100- or 150-grit papers or Dremel wheels (Figure 13-19), then 220, then 320 fine, then 500 super fine, and then tackle the micron-grade grits to eliminate sanding marks. Many 3D makers tend to

skimp on the earlier papers, to their detriment: these coarser grits are capable of stripping away the peaks of the layer lines. Go too fine too fast and you'll just round over the peaks without flattening them (see Figure 13-20).

Figure 13-19. *Dremel sanding and polishing kit*

Figure 13-20. *Rounding over the peaks*

After you've sanded a surface to your satisfaction, you can use a heat gun to gently warm the surface (Figure 13-21) until it melts slightly, which will erase many of the smaller scratches and restore the original printed color. Practice on scrap until you get the feel of it.

You can also can use the Novus plastic polishing system to get a remarkably smooth, polished surface on ABS prints, but most 3D modelers opt to just paint their prints instead.

Figure 13-21. *Use a heat gun to remove small scratches*

Matt Griffin is the Director of Community & Support at Adafruit Industries, a former Mak-erBot community manager, and author of the forthcoming MAKE book Design and Modeling for 3D Printing.

Weathering Your Prints

14.

Age 3D-printed objects to look like battered metal.

WRITTEN BY **JASON BABLER**

Weathering plastic or wood to make it look like aged metal (Figure 14-1) is an effect you can achieve in fewer steps than most people expect. The most basic weathering can be done with only two paints. Here, I'll show you how to weather and age using only three paints. This technique is how I weathered, in less than 10 minutes, a 3D-printed robot that we made here at the *MAKE* offices.

The Mega Make robot, designed by Make: Labs intern Dan Spangler, is comming soon as a downloadable, desktop-3D-printable project (http://makezine.com/go/mega-make).

Figure 14-1. *The aged metal look*

Here's what you'll need:

Metallic hobby paint
> I like Citadel's metallic paints; the color Ironbreaker (*http://www.games-workshop.com/gws/catalog/productDe tail.jsp?prodId=prod1500186a*) is my favorite.

Spray paint for the base coat
> I chose Krylon's Fusion brand.

Black acrylic paint
> Any kind will do.

Drybrushes
> The ones from MicroMark (*http://www.micromark.com/dry-brushes-set-of-4,7667.html*) work well. Alternately, you can take an old brush that has a lot of bristles and cut the end off so it's flat.

Other brushes
> Use a script brush or other thin brush suitable for details.

1. Pick a BaseCoat

The white piece shown in Figure 14-2 is actually from the back of the robot. It was 3D-printed here at the *MAKE* office. I painted the robot bright red, to mimic our Maker Faire robot mascot. I chose Krylon's Fusion brand, with the right red I was looking for. It dries pretty quickly and adheres to plastic really well. I skipped the priming process, but you can put this paint over primer easily enough.

Figure 14-2. *3D-printed part from the back of the robot*

2. Pick a Metallic Paint for Worn Edges

I like Citadel's metallic paints (Figure 14-3), but any hobby metallic paint will do. Citadel's Ironbreaker is my favorite, and it's quite brilliant.

Figure 14-3. *Citadel paints*

3. Start Drybrushing over the Piece

You need the right type of brush to get the right drybrushing results. Take an old brush that has a lot of bristles and cut the end off so it's flat, or buy a specific brush that's made for drybrushing, like MicroMark's Dry Brushes (Figure 14-4).

Figure 14-4. *Micro-Mark's Dry Brushes*

Dab the brush into the paint, and then remove most of the paint onto a paper towel. Drybrushing means exactly that: the brush should not be wet with paint. In fact, it needs to have barely any paint on it. Practice on something other than your model to get the hang of it. Drybrushing correctly will leave just a little paint on the raised edges of your piece. Remember to keep a light hand and go slowly (Figure 14-5).

Figure 14-5. *Drybrush the part*

For machined edges, I focus on adding a bit more paint around the main edges of the model. Imagine that those edges are what other objects come into contact with the most, and therefore need to look more worn (Figure 14-6).

Figure 14-6. *Emphasize worn edges*

4. Add Bigger Scrapes and Chipped Areas

Using a very thin script brush with few bristles, use the same paint and add bigger spots where the metal has "flacked" off even more (Figure 14-7). Don't go overboard with this;

using this technique everywhere will ruin the effect.

Figure 14-7. *Add bigger paint spots—sparingly*

5: Add Dirt and Grunge

Let's start adding some dirt and grime (Figure 14-8) to make this look dirty. We'll start by making a wash. A wash is pretty simple: just watered-down acrylic black paint. If you brush over newspaper and can still read the print, it's probably watered down enough.

Don't buy wash mediums at the art store. Instead, just add windshield washer fluid to thin acrylic paints. It dries quickly and will last you years.

Figure 14-8. *Use a wash to add "dirt"*

I generously apply the paint in all the nooks and crannies that dirt would normally gather in. If the paint is watery enough, it should run into all the valleys of your model quickly (Figure 14-9).

Figure 14-9. *Get in the nooks and crannies*

Depending on how much of this effect you want, you can dab up the paint immediately (Figure 14-10), leaving only a little black grime, or paint generously over the model and leave the paint on to dry for a more dramatic effect.

Figure 14-10. *Dab up the paint*

6. That's It!

It's pretty easy to get basic weathering effects, as you can see in Figure 14-11. There are more degrees of weathering you can expand upon: pigments, rusting agents, and other cool techniques are out there to find and invent!

Figure 14-11. *Finished weathered part*

Jason Babler is the Creative Director at MAKE Magazine and loves to sculpt.

PART VII

Applications

The Promise of 3D Printing

15

Printing the world on your desktop.

WRITTEN BY **STETT HOLBROOK**

It's a vision from a futuristic *Star Trek* universe: effortlessly creating three-dimensional objects on a machine in your home (or starship). And it's here today. The dam has now burst on the 3D printing market and this once out-of-reach technology is now available to just about anybody, for less than $1,000.

Will being able to print 3D objects on your desktop change the world?

Spend a few minutes talking to manufacturers of 3D printers or early adopters and you'll quickly hear them drop such heady adjectives as "game-changer," "disruptive," and "revolutionary."

An *Economist* article from April 2012 by Paul Markillie declared 3D printing and associated technologies nothing short of the "third industrial revolution."

"As manufacturing goes digital, a third great change is now gathering pace," he wrote. "The wheel is almost coming full circle, turning away from mass manufacturing and towards much more individualized production. And that in turn could bring some of the jobs back to rich countries that long ago lost them to the emerging world."

The personal computer, the printer, and the Internet made us all publishers. Now, with 3D printers, 3D scanners, and 3D design software, we can all be manufacturers as well.

Already companies are jockeying for position. High-end 3D printing pioneer 3D Systems bought competitor Z Corp. Two other big players—Objet and Stratasys—have merged. Industry darling MakerBot was named one of the top 20 startups in New York City and has been acquired as the desktop division of Stratasys.

3D printers, or machines that "print" three-dimensional, CAD-rendered objects by layering precisely extruded bits of molten plastic, resin, metal, and other materials, have actually been around since 1985—ironically, the same year the standard-setting HP LaserJet printer was introduced. The laser printer has become as commonplace as the personal computer. The same can't be said of the 3D printer. But that could be about to change.

Until recently, 3D printers were prohibitively expensive, less than user friendly, and hidden behind the doors of factories and R&D labs. But thanks to the innovative efforts of makers and the open source movement (which encourages freely sharing designs and software among enthusiasts), the price of the machines has reached the consumer level. Now, a growing community of makers, designers, and artists are embracing the technology and taking it in new directions. And you don't have to own one of the machines to use them—there are service providers that will do the printing for you. What you do with all this desktop manufacturing power is up to you.

To borrow a line (paraphrased from Karl Marx) from *Wired* Editor-in-Chief Chris Anderson's book, *Makers: The New Industrial Revolution*, power belongs to those who control the means of production. The power to manufacture a growing list of objects (toys, jewelry, spare parts, even prosthetic limbs) is now available to the masses—and the technology fits on your desktop.

"Global manufacturing can now work on any scale," Anderson states, "from one to millions. Customization and small batches are no longer impossible—in fact, they're the future."

He too sees revolution in the air: "The third industrial revolution is best seen as the combination of digital manufacturing and personal manufacturing: the industrialization of the maker movement."

Looking at the growing number of consumer-level 3D printers on the market has Anderson seeing 1983 all over again—the so-called "Mac moment" when Apple gave the masses a computer of their own: the Apple II. Apple didn't invent the computer, they just democratized it, Anderson notes. The same can be said of RepRap and Maker-Bot, two pioneers in the affordable consumer 3D printer market.

"A new class of users will produce a new class of uses," says Anderson. "I think it's historic."

Will 3D printers become as common in the home as DVD players and computers? Wonky software and documentation are the weak links now, but that will surely change. For now, Anderson tells parents this is the year to buy their kids a 3D printer for Christmas.

"They're not going to be quite sure what to do with it, but their kids will figure it out. That's the way big things start."

Dale Dougherty, founder and publisher of *MAKE*, isn't ready to pronounce the 3D printer a revolution just yet.

"I think we're at the very early stages, with hackers and early adopters figuring out what to do with it," he says. "It's opening new avenues for people who are creative and making things."

But the transformative potential is plain to see, Dougherty states. "It's Wal-Mart in the palm of your hand. That's the crazy promise of it."

Part of the excitement that surrounds 3D printing is the belief that now the barrier to entry has dropped; the genie is out of the bottle. Where this goes, nobody knows. "We live in a 3D world, but we currently create things in 2D," says Dougherty. What will it mean to have the means to live and create in the same dimension?

"We may go to a very different place."

Stett Holbrook is a senior editor at MAKE.

3D Printed Gallery 18

3D printing is being used in a wide array of customized applications, from practical objects to medicine to art. Here's a sampling of what's out there.

COMPILED BY ERIC CHU, ANNA KAZIUNAS FRANCE, GOLI MOHAMMADI, CRAIG COUDEN, AND THE EDITORS OF MAKE

Practical Objects

Print what you want, when you want it. These resourceful makers used their desktop 3D printers to overcome everyday problems by creating unique custom solutions that were just what they needed.

Pen Plotter Adapter

Miles Lightwood, Eagle Rock, CA
http://thingiverse.com/thing:7412

A friend is a vintage computer junkie and one of his recent purchases was a printer/plotter for his Sharp PC-1500A pocket computer. Although the printer works, the pens, being almost 30 years old, do not. Nor are they still available. So I took out the calipers, busted out OpenSCAD, fired up the MakerBot, and printed one out!

Figure 16-1. *Vintage pen plotter adapter*

Tackle Box Clip

Chris Krueger, Arlington Heights, IL

http://thingiverse.com/thing:9419

I bought a tackle box to help organize all my electrical components. Unfortunately, it was missing a clip. I fixed my problem by modeling a replacement from photos of the existing clip and caliper measurements. The replacement actually snaps tighter than the original, so I replaced them both!

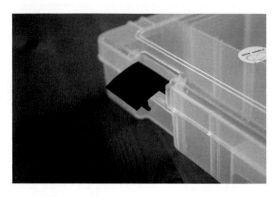

Figure 16-2. *Repaired tackle box*

Soda Tab Bender

Sean Michael Ragan, Austin, TX

http://thingiverse.com/thing:31635

I have this trick of making chain mail from soda can tabs. The tabs have to be bent first, and the mail looks best if they're all bent to exactly the same angle. So I designed and printed a custom jaw insert for an off-the-shelf pair of jeweler's pliers that bends the tabs exactly the same way each time.

Figure 16-3. *Chain mail in progress*

Router Collar Adapter

Bozo Cardozo, Ketchum, ID

http://thingiverse.com/thing:12648

My old Bosch 1611 is a killer router, but it's 20 years old and almost impossible to find parts for. I couldn't find a stock pattern collar adapter, so I just designed and printed my own.

Figure 16-4. *Adapting old tools*

Car Luggage Cover Fitting

Miguel Angelo de Oliveira, Hartsdale, NY

http://thingiverse.com/thing:10043

When some plastic part breaks in your car, the chances of finding a replacement and then being able to buy it without getting a second mortgage are slim to none. My Thing-O-Matic allowed me to replace a broken plastic fitting in my car's luggage cover, which saved me from having to buy a whole new one at roughly half the price of my 3D printer kit. I have now helped many others make vacuum cover clips, shower head holders, and many other hard-to-find replacement parts.

Figure 16-5. *Car luggage cover repair solution*

Car Window Crank

Michael Gregg, Palo Alto, CA

http://thingiverse.com/thing:20028

A good friend of mine was interested in 3D printing. We were hanging out one Sunday, and he broke the window crank in his Miata. The auto parts store was not open until Monday, so I whipped this up for him using SolidWorks and my MakerBot.

Figure 16-6. *Replacement window crank*

Bike Light Mount

Gian Pablo, San Francisco, California

http://thingiverse.com/thing:8947

I wanted to attach a rear light to my bike, but I have a rack installed, so it couldn't go on the seat post. The rack already had holes for a license plate, so I made this simple adapter that goes on the rack using the existing holes and has a spring clip that fits the bike light directly, so that I can attach and remove it quickly.

Figure 16-7. *Rear light attachment*

Ice Cream Maker Drill Attachment

Lee Holmes, Seattle, WA

http://thingiverse.com/thing:26835

I ordered a KitchenAid stand mixer and its ice cream maker attachment. The ice cream maker attachment arrived on a Monday, but the mixer itself wasn't going to arrive until the following Thursday. I wanted ice cream on Tuesday. This is my solution: an adapter to drive the ice cream maker with a standard 3/8 drill socket adapter (and a strong drill).

Figure 16-8. *Ice cream when you want it*

Doorbell Replacement

George Banovac, Stony Creek, Ontario

http://thingiverse.com/thing:10697

Some blockhead came along and broke my mother's doorbell by excessively pounding on it. Not just once, either. Instead of spending money on buying another doorbell button, I decided to make something stronger, better—not sure about the faster part, but you get the idea. I printed it in natural ABS. I can't even tell the difference from before; it looks exactly the same, but now it's stronger—a solid mini-brick instead of a hollow shell. In your face, doorbell murderers!

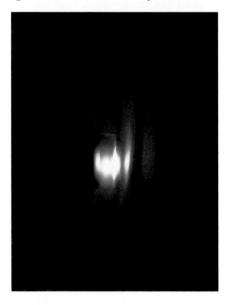

Figure 16-9. *Doorbell fix*

3D Products Now on the Market

All kinds of makers, artists, designers, and startups are taking advantage of the boom in 3D printing services to sell an amazing variety of custom-made or customizable objects.

Nervous System

http://n-e-r-v-o-u-s.com

Nervous System is an experimental design studio whose focus is on exploring algorithmic design. Taking inspiration from nature, they develop interactive tools that can be used to generate an infinite range of designs.

Currently available products include predesigned tableware, lamps, puzzles, and jewelry. Customers can also participate in the creation process by creating their own on-demand printed designs through the Radiolaria and Cell Cycle jewelry creation web-based applications.

Tofty's EDC Items

http://www.shapeways.com/shops/tofty

Specializing in useful every-day-carry (EDC) items, Tofty's multitools have many features and integrate prybars, bit drivers, nail-lifting tools, bottle openers, and metric hex drivers. He also sells flashlights and functional tritium lantern jewelry.

Figure 16-10. *Morph Bangle created with the Cell Cycle app and printed in sterling silver, available from Shapeways: http://shpws.me/pjLL*

Figure 16-11. *Single-piece prybar, available from Shapeways: http://shpws.me/llzP*

Continuum Fashion

http://continuumfashion.com

Continuum Fashion's Strvct nylon shoes are strong, lightweight, and made to order. After printing, they're fitted with a patent leather inner sole and coated with rubber on the bottom for traction. The designers also made headlines with their 3D-printed bikini. This fashion label also provides creation apps for their other on-demand custom fabric clothing lines CONSTRVCT, and the D.dress.

Figure 16-12. *3D-printed shoe from the "strvct" collection*

ModiBot

http://www.modibot.com

ModiBot is a figure-based build system that enables you to create your own toys by mixing and matching parts from the ever-expanding ModiBot part system. You can design and build your own fantastical creatures and characters by ordering the parts via Shapeways or downloading them and printing them yourself.

Figure 16-13. *ModiBot Angel at World Maker Faire*

Joaquin Baldwin

http://shpws.me/op7l

The award-winning creator of a number of short animated films, Joaquin Baldwin currently works at Disney Animation Studios by day and in his spare time designs 3D-printable objects that are "a bit of everything nerdy," from geeky gadgets to videogame-related sculptures and bizarre concept models, like his Bacon Mobius Strip or Caffeine Molecule Mug.

Figure 16-14. *Caffeine Molecule Mug, available from Shapeways: http://shpws.me/op7l*

Protos Eyewear

http://protoseyewear.com

San Francisco–based Protos Eyewear, creators of the 8-Bit Sunglasses, recently completed a successful crowdfunding campaign for custom 3D-printed eyewear. They have developed interactive software that enables the user to alter the shape of the glasses to flatter their facial structure using uploaded images.

Figure 16-15. *8-Bit Sunglasses*

Freakin' Sweet Knots

https://www.shapeways.com/shops/frea kinsweetknots

Created by a programmer who likes to tie knots, Freakin' Sweet Knots was born though John Allwine's experimental process of creating an engagement ring for his wife. During the process it occurred to me that I could write an app that could generate 3D models of similar rings so they could be 3D printed.

You can create your own knot ring through the Freakin' Sweet Knots app (*http://knots.freakinsweetapps.com*), which uploads directly to Shapeways.

Figure 16-16. *Joyce's Tapered Ring, available from Shapeways: http://shpws.me/olJo*

Polychemy

http://polychemy.com

Polychemy is a Singapore-based online designer boutique that makes a variety of 3D-printed products, from jewelry to mobile phone covers, and works with prominent 3D print artists to help sell their work.

Their iPhone cases are printed in a strong, flexible nylon plastic called polyamide, and you can have your name custom-printed into the case.

Figure 16-17. *iPhone case printed in red Polyamide plastic*

3D Printing in Medicine

From bioprinting to prosthetics, the medical community is embracing 3D printing in a multitude of innovative ways.

Revolutionary Replacements

http://makezine.com/go/wakeforest

Researchers at Wake Forest University's Institute for Regenerative Medicine in North Carolina have created a one-of-a-kind 3D printer designed to print replacement tissues and organs.

Data from CT or MRI scans is used to first create a 3D model. Living cells and biomaterials that hold the cells together are then printed into 3D shapes and implanted in the body, where they continue to develop using the body's natural regenerative processes. The team has successfully engineered a miniature kidney implanted in a steer and aims to 3D print similar versions for humans.

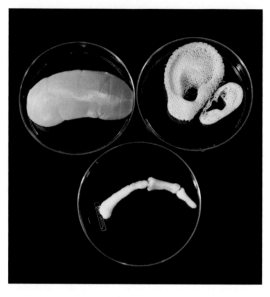

Figure 16-18. *3D printed replacement tissues and organs*

Beauty and the Beak

http://makezine.com/go/beauty

After a poacher in Alaska shot off her upper beak, Beauty the Bald Eagle was found emaciated, unable to eat, drink, or preen her feathers, her tongue and sinuses exposed. Despite recommendations that she be euthanized, raptor specialist Jane Cantwell of Birds of Prey Northwest refused to give up. She teamed up with Nate Calvin, founder of Kinetic Engineering Group. Despite having no previous prosthetics experience, Calvin designed a replacement by making a mold of the remaining beak, scanning it, modifying it in SolidWorks, and creating a temporary 3D printed beak in a nylon composite. A titanium baseplate was affixed to Beauty's remaining upper beak to act as a guide for the final printed titanium beak.

Figure 16-19. *Beauty the Bald Eagle—before and after her titanium 3D printed beak implant.*

Magic Arms

http://makezine.com/go/stratasys

Using hinged metal bars and resistive rubber bands, the Wilmington Robotic Exoskeleton (WREX) gives patients with underdeveloped arms a wide range of arm motion.

The original WREX, made from machined parts, fit children as young as six, but for two-year-old Emma Lavelle, researchers discovered they could 3D print smaller, lighter parts. Printed using a Stratasys Dimension 3D printer, the ABS plastic exoskeleton allows for easy customization and fine-tuning for the 15 children now using it.

Figure 16-20. *WREX Magic Arms*

Tissue Engineering

http://organovo.com

San Diego-based innovators Organovo have engineered several custom, commercial NovoGen MMX Bioprinters, capable of printing tissue structures.

The machines feature dual extruders: one that prints a water-soluble gel that acts as the scaffolding, and another that fills the armature with a bio-ink of living cells (each drop containing 10K to 30K cells) that naturally flow together and fuse. The fundamental nature of biological materials is to self-organize, and after an incubation period, the cells continue to develop and grow without the gel component. The team has been able to print a 1 mm-diameter, 5 cm-long blood vessel in 30 minutes.

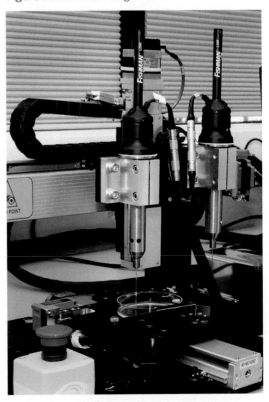

Figure 16-21. *Organovo's NovoGen MMX Bioprinter.*

Designer Prosthetics

http://bespokeinnovations.com

Bucking the typical one-style-suites-all model of conventional prosthetics, San Francisco-based Bespoke Innovations makes custom coverings, called Fairings, that surround an existing prosthetic leg, recreating the user's natural leg shape through 3D scanning, designing, and printing.

Conceived by industrial designer Scott Summit and orthopedic surgeon Dr. Kenneth Trauner, Bespoke Fairings are designed in collaboration with the patient, aiming to reflect the individual by offering a number of material and pattern options, including leather, chrome plating, and tattoos.

Figure 16-22. *Bespoke Fairings prosthetic leg covering.*

Jaws of Innovation

http://makezine.com/go/jaw

A Dutch woman in her 80s was fitted with a 3D-printed jawbone after having her infected mandible removed. Belgian company LayerWise modeled the replacement from a CT scan of the original bone.

Using selective laser sintering, the replacement was made of titanium and covered with a bioceramic coating. The replacement weighed just over an ounce more than the original, and the woman was reportedly speaking within hours after surgery.

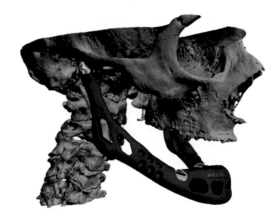

Figure 16-23. *3D-printed jawbone*

Not Your Average Chem Set

http://makezine.com/go/reactionware
Researchers at the University of Glasgow have developed a way to synthesize custom labware on a much smaller scale. Using a low-cost 3D printer and open source CAD software, researchers printed customized vessels, called "reactionware," out of polymer gel laced with chemical reagents that foster chemical reactions (think of an Erlenmeyer flask infused with chemicals for that day's lab experiment). It could eventually lead to pharmacists making small, customized batches of medicines for individual patients.

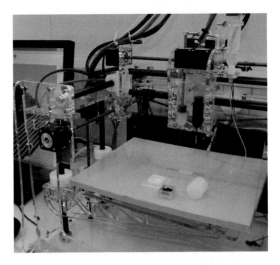

Figure 16-24. *Printing reactionware*

Novel and Artistic Prints

These creative makers created some of the most unique and memorable 3D printed objects to date and most of them are printable on a desktop 3D printer.

Orihon (Accordion Book)

Tom Burtonwood
http://www.thingiverse.com/thing:110411
A book of textures and reliefs made up of six 3D scans of works from the MET, the Art Institute of Chicago, the American Museum of Natural History, and the Field Museum of Natural History. Inspired by a call for submissions from the Center for Book and Paper Arts at Columbia College that called for both print-on-demand and photographic books.

Figure 16-25. *Orihon*

InMoov Humanoid Robot

Gael Langevin "hairygael"

http://thingiverse.com/hairygael

InMoov is the open source life-sized humanoid robot that you can print on a desktop 3D printer (!) and animate. In addition to the printable files, Gael also provides assembly instructions and diagrams to help you get started: *http://www.inmoov.fr*

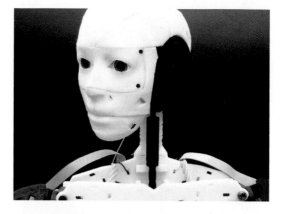

Figure 16-26. *Life size humanoid robot*

The Makerlele - MK1

Brent J. Rosenburgh "ErikJDurwoodII"

http://www.thingiverse.com/thing:34363

A fully desktop 3D printable (and playable) ukulele that uses an acoustic transducer focus and projects the sound through the body to create a fuller tone. Just print, add bolts and strings, tune it up, and take it for a spin.

Figure 16-27. *Makerlele*

Head of a Horse of Selene

Cosmo Wenman

http://www.thingiverse.com/thing:32228

http://cosmowenman.com

This work is a scanned recreation of the famous Parthenon sculpture of the horses who drew the chariot of Selene, the moon goddess. Cosmo's recreation of the Head of a Horse of Selene was printed in PLA on a MakerBot Replicator 3D printer and is the same size and scale as the original sculpture. Finished in "Epic Bronze" using his Alternate Reality Patinas, it is a new take on a beloved work of antiquity.

Figure 16-28. *Head of a Horse of Selene*

Orbicular Lamp Series

Nervous System

http://n-e-r-v-o-u-s.com

Part of the Hyphae lighting collection from Nervous System, the Orbicular lamp is based on how veins form in leaves. These complex and unconventional geometries are created using a novel computer simulation process. Each lamp in the collection is a one-of-a-kind design 3D printed in nylon plastic. Lit by eco-friendly LEDs, they cast dramatic shadows on the walls and ceiling.

Figure 16-29. *Orbicular lamp*

Automatic Transmission Model

Emmett Lalish "emmett"

http://www.thingiverse.com/thing:34778

Have you ever wondered how an automatic transmission works? Emmett Lalish did, so he found out and designed this working desktop model. With six forward speeds (and one reverse speed), this model is a great teaching tool.

Figure 16-30. *Working automatic transmission model*

Marble Run

Adam Fontenault and Chris Boynton

http://little-badger.com/portfolio/makerbot-marble-run

This marble run has over 2,000 individual pieces printed from ABS snapped together and has five separate forking paths that carry each marble on its journey. There are two of these colossal displays, one at the MakerBot NYC store and one in the MakerBot offices.

Figure 16-31. *MakerBot Marble Run*

Dita's Gown

Michael Schmidt and Francis Bitonti

*http://www.michaelschmidtstu
dios.com/dita-von-teese.html*

http://francisbitonti.com/Dita-s-Gown

A fully articulated, completely customized 3D-printed gown based on the Fibonacci sequence, designed to be worn by Dita Von Teese, the queen of burlesque. The dress was designed by Michael Schmidt, 3D modeled by architect Francis Bitonti, and printed in nylon by Shapeways. After printing, the garment was assembled from 17 different pieces, dyed black, and lacquered, and over 13,000 Swarovski crystals were attached to create a "glowing" effect.

Figure 16-32. *Dita's Gown - Albert Sanchez Photography*

Joy Division's *Unknown Pleasures* Cover

Michael Zoellner aka "emnullfuenf"

http://www.thingiverse.com/thing:92971

A printable model of Joy Division's iconic *Unknown Pleasures* cover representing pulsar PSR B1919+21 waveforms. Unable to find a vector graphic or 3D model of the cover art, Michael Zoellner ended up hand-tracing the waves, exporting as a DXF, and then extruding them in OpenSCAD.

Figure 16-33. *3D printed Unknown Pleasures cover*

Digital Grotesque

Michael Hansmeyer and Benjamin Dillenburger

http://www.digital-grotesque.com
http://www.michael-hansmeyer.com
http://benjamin-dillenburger.com

Digital Grotesque is an elaborate room designed through customized algorithms that is entirely 3D printed out of sandstone. The sandstone was infiltrated with resin to increase the structural stability and close the pores, then coated with pigment, alcohol, and shellac.

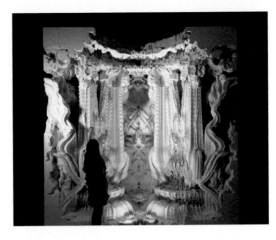

Figure 16-34. *Digital Grotesque*

Eric Chu is a MAKE Labs Alumnus, an engineering student, yo-yo hacker, robot builder, and fried rice aficionado.

Anna Kaziunas France is the Digital Fabrication Editor at Maker Media.

Goli Mohammadi is a Senior Editor at MAKE who has worked on MAKE magazine since the first issue.

Craig Couden is an Editorial Assistant with Maker Media.

Dream Machine | 17

Dissolving the boundaries between imagination and physical reality.

WRITTEN BY **KEVIN MACK**

PHOTOGRAPHY BY **KEVIN MACK**

As an artist, I've always been driven to create art that no one has ever seen before. My search for the innovative led me to computers in the early '80s. I recognized the potential for digital technology in art and immersed myself.

A career in visual effects put me at the cutting edge of this technology, where I could explore and experiment with digital 3D tools like procedural modeling and volumetric sculpting, and apply them in unintended ways. I learned to build brand-new tools and processes that made no attempt to emulate traditional methods. These processes resul-

ted in unique virtual artifacts that imbued my digital objects with a surreal and mysterious quality. They were new.

As these newfound capabilities enabled my wildest art ambitions, along came 3D printing, which made it possible to manifest my digital creations in the physical world. It was like a science fiction dream come true.

One reason I find 3D printing so compelling is that it enables production of objects that could not be made in any other way. Rapid prototyping lacks many of the limitations of traditional object-creation methods. By exploiting these capabilities, I'm able to add a powerful dimension to my sculptures: not only are they new, but they appear to be impossible.

Humans are immersed in the world of manufactured objects, and our brains have an intuitive understanding of how things are made. We've come to expect certain artifacts of the fabrication processes, such as seams, limited part complexity, and lack of compound curvature. 3D printing enables the creation of objects that conspicuously defy these expectations. When you experience these objects and hold them in your hands, there is an immediate intuitive awareness that you are seeing something that wasn't possible before. This can be a profoundly surreal experience.

My dream of creating art that no one has ever seen before has come true. Digital technology and 3D printing open up a vast space of entirely new aesthetic possibilities. I am creating and discovering unimaginable forms and sculptures that couldn't exist before.

This is just the beginning. Technology is constantly advancing. In the future, I predict that creative technologies will completely dis-

solve the boundaries between imagination and physical reality. I don't know if the world is ready, but I can't wait.

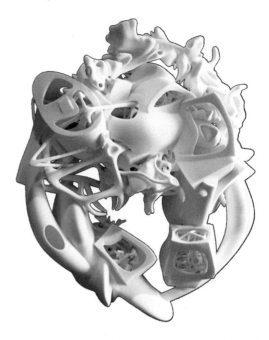

Kevin Mack (http://kevinmackart.com) is a pioneering digital artist and Academy Award-winning visual effects designer. Mack uses science and technology to make art that depicts and inspires the transcendent.

Desert Manufacturer

18

Markus Kayser's sand and sun 3D printer.

WRITTEN BY **LAURA KINIRY**

Think that deserts lack resources? Not to Markus Kayser. The MIT research assistant with a master's in product design has created a 3D printer that can make glass objects using a desert's two most abundant resources: sun and sand. "I began asking myself," says Kayser, "what if I could build a machine which would act as a kind of translator between the two?" His Solar Sinter (*http://markuskayser.com/work/solarsinter*) (Figure 18-1) is the direct result.

Figure 18-1. *The Solar Sinter*

Based on a type of 3D printing known as selective laser sintering (SLS), Kayser's Solar Sinter uses the sun's rays as a laser and sand rather than resins to create exact physical replicas of his digital designs. The printer includes a large Fresnel lens that's always facing the sun (by way of an electronic sun-tracking device), stepper motors to move and load its sandbox, and two 60-watt photovoltaic panels that provide electricity to charge the battery that drives the motors and electronics of the machine.

His first time out with Solar Sinter, Kayser produced both a bowl and a tile, as well as a sculptural piece. "Once I input the design I'm trying to produce via an SD card," he says, "the machine reads its code and then moves the sandbox along to the correct x and y coordinates at 1 mm per second, while the lens focuses a light beam that produces temperatures between 1,400°C and 1,600°C, more than enough to melt the sand." The objects (Figure 18-2) are built layer by layer over the course of several hours.

Figure 18-2. *Sintered objects*

"[In the future] printing directly onto the desert floor with multiple lenses melting sand into walls and eventually building architecture in desert environments could be a real prospect," Kayser muses.

Laura Kiniry is a San Francisco-based freelance writer and regular MAKE contributor.

How I Printed a Humanoid

WRITTEN BY **MICHAEL OVERSTREET**

In recent years I've experimented with 3D printing the structural brackets for my humanoid robot Boomer (Figure 19-1). Boomer and I compete at RoboGames and show off at Maker Faires, and people always ask me: Why 3D printing? Why not just make the brackets out of metal like everyone else?

Figure 19-1. *Michael and Boomer*

I started to do it just because I could. The advent of cheap DIY 3D printers has given makers a new way to manufacture and customize objects, and I was simply exploring this new process. As I printed more parts for my robot, I realized that it could be done—and done cheaply.

Then I saw the DARwIn-OP at the 2010 International Conference on Humanoid Robots. After seeing how capable and ground-breaking it was, I wanted one. But how could I afford it? A new DARwIn-OP from Robotis costs $12,000.

Humanoid Evolution

The DARwIn-OP (Dynamic Anthropomorphic Robot with Intelligence—Open Platform) is a state-of-the-art research and development humanoid robot created by Virginia Tech's Robotics and Mechanisms Laboratory (RoMeLa), led by Dr. Dennis Hong, in collaboration with the University of Pennsylvania, Purdue University, and the South Korean company Robotis, with support from the National Science Foundation.

Weighing in at just 2.9 kg and standing 45.5 cm tall, the DARwIn-OP won the gold medal in the autonomous RoboCup Soccer Humanoid League, Kid Size class, in both 2011 and 2012.

That's how this project got started. My mission was to print as much of the robot as possible using a DIY 3D printer costing $2,000 or less. DARwIn-OP is an open hardware and software project, so all of the 3D files and plans are free online. I bought 20 Dynamixel MX-28T servos and the complete electronics kit from Robotis.

Today, my DARwIn-OP clone (Figure 19-2) is fully assembled, and I've spent about $6,100, not counting the costs of the 3D printers I used. That's still a lot—but it's almost half off the factory price.

Figure 19-2. *Meet the clone!*

Here are some of the things that made this possible:

90% DIY printed

I print my parts in ABS plastic, as PLA is weaker and more brittle. I strongly agree with the review of the UP! Plus/Afinia printer in the 2012 *Make: Ultimate Guide to 3D Printing*. It's one of the best on the market, and it has printed 90% of my clone: servo brackets, structural framing, and body covers.

Top servos

Robotis Dynamixel digital servomotors are the leading robotics servos in the

world—fast, high torque, and very high resolution.

Glues not screws

Nuts and bolts are not ideal for plastic. I glue my 3D-printed brackets together using Micro-Mark's Same Stuff (Figure 19-3) liquid plastic welder or an ABS-acetone slurry.

Figure 19-3. *Micro-Mark's Same Stuff*

Off-the-shelf brains

The DARwIn-OP is controlled by an affordable Fit-PC2 compact PC and a Robotis CM-730 servo controller (Figure 19-4).

Figure 19-4. *The brains of the bot*

Here are some of the challenges I faced:

The toughest parts

A few parts, designed for aluminum or injection-molded plastic, are almost impossible for a DIY printer. The front body cover must be printed at the highest resolution—that's 16 hours on my printer, and a lot can go wrong in 16 hours. I had Shapeways (*http://shapeways.com*) print the head and body covers, using their laser sintering process. I hope to modify these parts for a DIY printer—or to find one that can handle them.

Tricky widths

Some parts are one and a half or two and a half extruder paths wide instead of two or three. Many DIY printers and slicer programs can't overcome this problem, causing hollow passageways within walls that weaken the part. Also prob-

lematic: body covers are barely one path wide.

Lessons Learned

Here are a few more things that helped along the way.

Free 3D design

I had to fine-tune some of the parts, so I used Autodesk's 123D design software to create my STL files, both for its price (free) and its ease of use (see Figure 19-5).

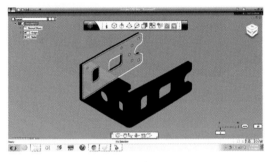

Figure 19-5. *3D-printed bracket*

Rafts and support material

I like the UP! Plus 3D printer's accuracy and its automatic generation of both rafts (disposable footings that prevent warping) and support material to prop up overhangs, bridges, and screw holes (Figure 19-6). I still had to play around with the orientation of parts on the build platform, because this can affect the printing of overhangs and supports (see Figures 19-7 and 19-8).

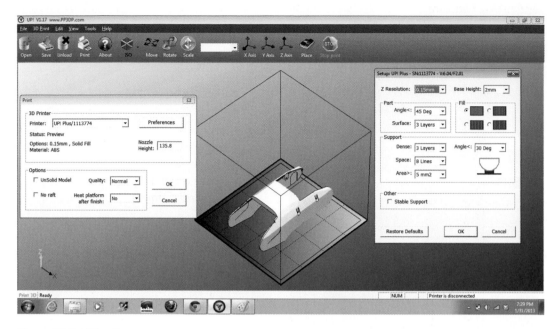

Figure 19-6. *UP! software*

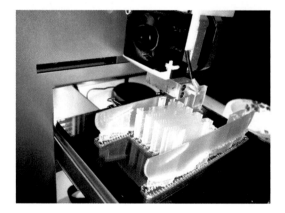

Figure 19-7. *Printing with support material on the UP! Plus*

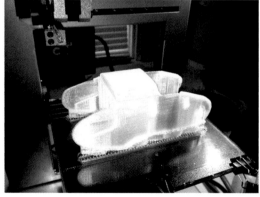

Figure 19-8. *Finished printed humanoid part*

Heated platform

> Printing in ABS requires a build platform heated to about 110°F. To improve performance, I upgraded my Up Plus platform to glass covered with Kapton tape for adhesion.

Leveling the platform

Learn the paper leveling trick: if a sheet of paper can freely move between the extruder tip and the build platform, you're OK. If not, then the extruder tip is too close. If it moves too easily or you see a gap, then it's too far away.

Awesome manuals

The open source DARwIn-OP manuals are the most comprehensive and detailed I've ever worked with—they break down the entire assembly into easy steps (Figures 19-9 through 19-11). I thank Dr. J.K. Han of Robotis for the tireless work he must have put into creating them!

Figure 19-10. *Humanoid robot design from the Robotis assembly manual*

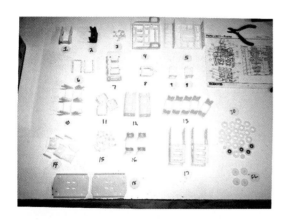

Figure 19-9. *Step-by-step humanoid documentation*

Figure 19-11. *Another humanoid robot design from the Robotis assembly manuals*

DIY vs. Commercial

To print the whole robot, I used two spools of plastic that cost about $90 total. Shapeways's price is almost $1,000. That's close to the price of many DIY 3D printers, and in my opinion the quality of DIY prints is now 70% to 80% as good as that of professional prints. Still, I'm considering having all the covers printed by Shapeways for about $400, as they're the most difficult parts to get right.

What's Next

Next I'll download the control software to the Fit-PC2 and get this robot to walk, talk, and see, with as few modifications to the software and hardware as possible.

I may need to redesign some of the brackets and frames, because these were designed for aluminum. My friend Yoshihiro Shibata is helping me analyze DARwIn-OP's structure to find its weak points when printed in plastic (see Figures 19-12 and 19-13). My plan is to redesign only the parts that break, to make them stronger, as I don't want to add unnecessary weight.

Figure 19-12. *Analyzing DARwIn-OP's structure*

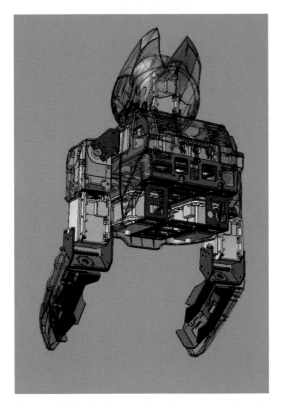

Figure 19-13. *Finding weak points when printed in plastic*

That's one of the great benefits of personal manufacturing: if something doesn't work, just redesign it and print it out. This process is called *iterative design and experimentation*. You can do it as many times as needed because the cost of a part is only a few dollars and you just have to wait for it to be printed!

After the robot is fully functional, I'll modify its covers to resemble my favorite fictional humanoids, Robby, Gort, Atom, and C-3PO, and maybe add a fully articulating hand with fingers.

But I've come to realize that I may never be fully done with this project. I think the only limit is my imagination. This, I think, is the greatest benefit that we'll get from the personal fabrication movement. It has opened up limitless possibilities for individuals to fully explore in solid form their dreams, ideas, and imaginations.

Special thanks to Luis Rodriguez, Rob Giseburt, Paul Piong, Roc Terrell, James Rao, and Kayla Kim for their printers and knowledge.

Michael Overstreet (http://mike-ibioloid.blogspot.com) is a computer programmer by day and amateur roboticist by night. He and his humanoid robot Boomer have won medals in the last six RoboGames. He's a founding member of the Cowtown Computer Congress hackerspace in Kansas City, MO., and has attended all of the national Maker Faires.

PART VIII

Other Ways to Make 3D Objects

Milling 3D Objects

20

The takeaway on subtractive vs. additive fabrication.

WRITTEN BY **TOM OWAD**

There's a lot of well-deserved excitement surrounding 3D printers, and for the avid DIYer much of it is focused on their ability to "self-replicate" by making their own parts. But is a 3D printer the right tool for you?

A 3D printer's fabrication technique is additive—most of them use a hot plastic extruder to "print" a plastic model. This contrasts with subtractive fabrication tools, which start with a solid block of material and use a cutter to remove the excess. Subtractive fabrication is far more common than additive, especially when working with metal and wood. Lathes, mills, saws, drills, and other CNCs like laser and vinyl cutters are all subtractive tools.

The Subtractive Equivalent of 3DP

A CNC milling machine or router is the subtractive equivalent of the 3D printer. For the hobbyist, milling is inferior to printing in numerous ways:

- Milling inherently causes waste, and without some sort of dust control, that waste gets flung throughout the room.

- Milling is more dangerous—while it's possible that a plastic extruder might overheat and catch fire, I've already had a (minor) fire with my CNC router, and there's the added danger of a blade spinning at 20,000 rpm sending bits of itself, or even your workpiece, flying at you.

- A mill or router is necessarily larger and heavier than a 3D printer and consequently more expensive and more difficult to move. It requires a positioning system that can maintain accuracy when encountering resistance, and motors powerful enough to drive it.

- Software preparation is also more complex for milling. After drawing the object you wish to make in a CAD or 3D modeling program, it's necessary to generate toolpaths with computer-aided manufacturing (CAM) software. This involves specifying the dimensions and location of the stock material, the dimensions and characteristics of the end mill (cutter), and speeds for the axes and spindle. The tools to do this tend to be complex, and a bit daunting for the first-time user.

From the user's perspective, CNC milling is a much more complex process than printing is. CNC milling does, however, have a significant advantage over 3D printing: the technology is mature. Home 3D printers are improving at a tremendous rate, but there's often still a lot of tinkering and experimentation involved in getting a good print.

What Do You Want to Make?

If your interests tend toward larger and more structural creations, go with milling. Also consider that it's cheaper to work in wood than in plastic, and that you're likely to get substantially superior results.

On the other hand, making complex 3D objects is a lot more complicated with CNC milling than with a 3D printer. There are free tools for doing 2.5D milling, but CAM software for 3D milling can be very expensive and difficult to use.

Doing 3D work for 3D printing is much easier. You can design your models in a free program like SketchUp or Inventor Fusion, and then export an STL file. Slicing software converts the STL file automatically to toolpaths in G-code, then sends the G-code directly to your 3D printer. With 3D printing, there's no need to tweak cutting paths, and no worry about the tool crashing into your work.

If creating small 3D objects is your goal, a 3D printer is the right choice.

DIY CNC

Going with milling doesn't mean you have to give up on self-replication or on making your own machine. Patrick Hood-Daniel, author of *Build Your Own CNC Machine*, makes a scratch-build CNC kit capable of making all of its custom parts, just like the RepRap 3D printer and its progeny. The frame is built of custom-cut plywood. Everything else is standard hardware.

The aluminum angle, bolts, and screws are available from any hardware store. The leadscrews and anti-backlash nuts will probably have to be mail-ordered from McMaster-Carr and DumpsterCNC, but you can get by with lesser hardware store parts in a pinch. The stepper motors and stepper drivers are completely generic and available from countless sources. The spindle is an ordinary wood router. I use a Porter-Cable 892.

As with the RepRap, the trick with Patrick's CNC router is getting a seed unit. Fortunately, any CNC router that can handle a 2'×4' sheet of plywood is capable of making the parts. Check out your local hackerspace, TechShop, Fab Lab and even the MAKE forums or the Home Shop Machinist, and you're likely to find somebody local with the necessary equipment. (If you can't find anybody to cut the parts for you, Patrick sells several kits.)

Plans Avaliable from Buildyourcnc.com

To get the CAD and CAM files for the parts, go to *http://buildyourcnc.com/cnckit2.aspx* and download the plans for the CNC Routing Machine Kit Version 1.3. Videos on the

Tom Owad is a Macintosh consultant in York, PA, and Editor of Applefritter (http://applefritter.com). He is the author of Apple I Replica Creation.

website explain how to put everything together. The files are licensed under the Creative Commons Attribution-NonCommercial license, which means you can make machines for yourself and for your friends, but you can't sell them commercially.

The CAM files are in the proprietary CamBam format, but there's a free version available, and it can read and write DXF files. Most good CAM software is very expensive, so if you don't already have a favorite, stick with CamBam. It can do both CAD and CAM, and even the free version is tremendously full-featured.

Another exciting new option is the MakerSlide system, available from Inventables (*http://inventables.com*). Perhaps the most significant weakness in the BuildYourCNC design is its decreased rigidity and precision due to its wooden design. MakerSlide integrates V-rails into what is otherwise a standard aluminum extrusion system (available from 80/20, Misumi, and others). By building your machine out of a combination of MakerSlides and stock aluminum extrusion, it's possible to create a more rigid, precise machine than you could with wood.

Even More Mills

This chapter was originally published in *Make: Ultimate Guide to 3D Printing* (*http://makezine.com/volume/make-ultimate-guide-to-3d-printing/*) (2012). Since then there have been several new desktop CNCs and DIY linear motion systems released at a low price point.

> *It should also be noted that there are many different types of subtractive manufacturing in addition to milling, including laser cutters, vinyl cutters, and various CNC drag knife attachments.*

Othermill

http://otherfab.com

The Othermill is a small-scale, high-precision CNC mill from Otherfab that evolved from the MTM Snap project (Machines That Make) created by Jonathan Ward at the MIT Center of Bits and Atoms (*http://mtm.cba.mit.edu/*). Jonathan now works for Otherfab.

This machine has a working area of 5.5"x4.5"x1.4" and is optimized for milling circuit boards out of FR-1 copper-clad PCB stock using a quiet, high-speed spindle. It can also machine soft woods, machinable wax, plastics and nonferrous metals. Unlike CNCs made from plywood, MDF, and other materials that are prone to warping, the Othermill is made from chemical resistant and moisture-proof HDPE.

In addition, the team at Othermill has taken on the task of both simplifying and beautifying the CAM process for milling by creating their own custom CAM package for the Othermill, which at the time of this writing is still in development. The cost of the Othermill will be a little over $1,000 for a fully assembled and ready-to-use machine.

Shapeoko 2

- Documentation: *http://www.shapeoko.com*
- Available though Inventables: *https://www.inventables.com/technologies/desktop-cnc-mill-kits-shapeoko*

Shapeoko is an open source CNC milling machine kit designed by Edward Ford that can be built in a weekend, depending on your level of expertise. It is available from Inventables as a Mechanical Kit ($299) or a Full Kit ($649 for 110V / $685.00 for 220V).

The Shapeoko lives up to its open source billing and provides full documentation for all of the machine parts, including the CAD files. It also uses the MakerSlide system discussed earlier in this chapter. The machine is driven by an open source hardware stack and currently uses grbl as the G-code interpreter and G-code sender (Windows) or gctrl on Mac and Linux. Materials for milling include plastics, woods, and nonferrous metals like aluminum and brass. However, it should be noted that Shapeoko strongly recommends a spindle upgrade (instead of using a rotary tool) if you plan to machine metals.

Linear Motion Systems

In addition to the complete and kit CNC milling options already mentioned, there are several companies providing open source aluminum extrusion combined with linear motion systems which make it easy to build your own custom CNC; MakerSlide, OpenBeam and OpenBuilds.

MakerSlide

- Documentation: *http://www.maker slide.com*
- Available though Inventables: *https://www.inventables.com/technologies/makerslide*

MakerSlide was created by Bart Dring and implements the traditional V wheel running on a V rail system of linear motion commonly used in commercial CNC machines. It was the first popular, affordable, and open source system of it's kind and is sold through Inventables.com. The Shapeoeko 2 uses MakerSlide.

OpenBeam

- Documentation: *http://blog.openbeamusa.com*
- Avaliable through Amazon: *http://store.openbeamusa.com*

OpenBeam open source aluminum extrusions are used in the Mini Kossel and Kossel Pro 3D printers, which are both open source RepRap printers, as well as commercial (RepStrap) 3D printers that have emerged from a partnership between the Rostock and Mini Kossel creator Johann Rocholl and Terence Tam of OpenBeam.

OpenBuilds

- Documentation: *http://www.open builds.com*
- Avaliable through OpenBuilds Part Store: *http://openbuildspartstore.com*

The OpenBuilds Part Store sells OpenRail open source aluminum V-slot extrusions and other linear motion components. OpenBuilds V-slots are used in the Bukito 3D printer from Deezmaker.

—Anna Kaziunas France

White Chocolate Skulls in PLA Trays

21

Making molds for chocolate casting with 3D printing.

WRITTEN BY **ANNA KAZIUNAS FRANCE**

Every Halloween I make treats to give away. I don a costume and distribute them to everyone I meet during my Halloween travels. I call it "reverse trick-o-treating." During October of 2012, I created a 3D-printed chocolate mold maker and individual trays in which to place the chocolates before bagging them. To cast the chocolates, I used food-safe silicone to make the final chocolate mold (Figure 21-1) from the 3D-printed mold maker.

Figure 21-1. *The white chocolate skulls*

I had scanned the skull a few months earlier using 123D Catch and had already created a 3D-printed necklace with it for my Kali Halloween costume, which I wore (Figure 21-2) when I handed out the chocolates.

Figure 21-2. *Happy Halloween 2012 from Kali and Finn & Jake!*

I used an OpenSCAD (*http://openscad.org*) script to make a mashup of a parametric box (by Thingiverse user *acker*) and my skull to create the mold maker. The trays for the chocolates were created using the same parametric box script.

The candy trays were printed in PLA, and were only used for presenting (not molding) the food. If you're worried, use a layer of parchment or wax paper to separate the skulls from the tray. I cast many batches of white skull chocolates and then placed them in the PLA trays. I placed the trays in treat bags and sealed them with twist ties.

Here are the files you'll need:

The Chocolate Skull Mold Maker
http://www.thingiverse.com/thing:33432

Skull with Pointed Teeth (cleaned and repaired scan)
http://www.thingiverse.com/thing:31998

The original skull scan (before cleanup) is available from 123D Catch
http://www.123dapp.com/obj-Catch/
Skull-with-Pointed-Teeth/859975

Bill of Materials

You will need the following items in order to recreate the White Skull Chocolates. My 3D printing supplies are shown in Figure 21-4, and you can see the others in Figure 21-3.

- Access to a 3D printer (I used a MakerBot Replicator) or use of a 3D printing service (like Ponoko or Shapeways)

- PLA filament to print on that printer

- Smooth On Smooth-Sil 940 (*http://www.smooth-on.com/a25/Smooth-Sil%3D-940-Suitable-For-Food-Related-Applications/article_info.html*) food-safe silicone rubber

- Chocolate Melts (no tempering required). I like Chocoley's Bada Bing Bada Boom Candy & Molding Formula (*http://www.chocoley.com/badabingbadaboom/candyandmolding.htm*)

- Soy Lecithin (softgels. I found these at Whole Foods in the supplement aisle

- Food thermometer. I find that digital ones with big letters are easiest to read; I purchased mine at my local Harbor Freight (*http://www.harborfreight.com/instant-read-digital-thermometer-95382.html*) for wicked cheap.

- Double boiler (or a tall soup pot and a small glass mixing bowl)

- Sharp knife

- Cutting board

- Small spatula/spreader

Optional items:

- Cocoa powder (I used Green & Blacks (*http://www.greenandblacks.com/ca/what-we-make/home-baking/cocoa-powder.html*))

- Candymaker's cotton gloves (*http://www.chocoley.com/supplies.htm*) (keep your fingers from marring the chocolate when demolding)

- Food-safe squeeze bottles or a chocolate funnel (keeps the process from becoming overly messy)

- Bamboo steamer (can be used in place of a double boiler or tall soup pot/glass bowl combo)

- Plastic treat bags with twist ties (I bought mine at Joann's Fabric)

Figure 21-3. *My chocolate casting supplies*

Figure 21-4. *My 3D printing supplies*

1. Print the Mold Maker on a 3D Printer

Print the mold form on a 3D printer (Figure 21-5). For people printing on their 3D printers at home, I recommend printing the Chocolate Skull Mold Maker with 13% infill and three shells. I have provided two versions of the Chocolate Mold Maker STL file, one with thin walls (1.3 mm) and one with

thicker walls (2.3 mm). The one with thin walls is not completely watertight if you print it with a raft. I used liquid tape on the bottom to make it water tight, but in the end the rubber mold mix was very thick and I don't think the liquid tape was actually necessary.

Figure 21-5. *Freshly 3D-printed mold maker*

The thicker one will take much longer to print. The thin-walled one should be fine to pour the silicone rubber into.

Design Your Own Mold in OpenSCAD

If you want to design your own mold instead of using the skull mold file included with the Chocolate Skull Mold Maker (*http://www.thingiverse.com/thing: 33432*), you can edit *skullCandyMold.scad* to use your own object.

I designed this mold using a mashup of an OpenSCAD parametric box (by Thingiverse user *acker*) and my scanned skull.

Before you make any changes, I suggest that you play around with the mold code as is. For that to work, make sure the *vampireSkull_0.2.stl* is located in the same directory as the OpenSCAD file.

Once you're sure it's working, you can substitute your own STL for the mold: just change the "filename" variable to the name of your STL. Make sure your STL is in the same directory as the *.scad* file.

Here's a render of the mold maker in OpenSCAD:

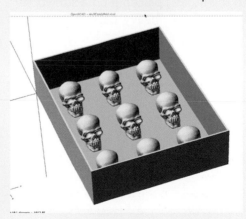

2. Mix and Pour the Smooth-Sil 940

Take the Smooth On Smooth-Sil 940 food-safe silicone rubber (Figure 21-6) and mix it according to the proportions on the package. Pour it into the 3D-printed chocolate mold maker. Fill the mold so that it covers the skulls by about a half inch, or close to the top of the mold.

Figure 21-6. *Ready to pour!*

I don't have access to a vacuum degasser (yet), so to help prevent bubbles from forming in the mold, I placed it on a subwoofer while music with heavy bass was playing (Figure 21-7). I also used an electric toothbrush without the brush head attached to vibrate the bottom and sides of the outside of the mold to get bubbles to come to the top.

Figure 21-7. *Alleviating bubbles with a subwoofer*

I don't know how effective these measures are, but I did not have any problems with bubbles or the mold material losing detail. Let the poured mold set for 24 hours and follow the heat curing instructions. Read the datasheet (*http://www.smooth-on.com/tb/files/Food_Grade_SS940.pdf*).

I covered the bottom of the thinner walled-mold with black liquid tape (liquid rubber), as shown in Figure 21-8. I was worried that the model was not completely watertight, but I don't think this is actually necessary. The Smooth-Sil 940 was so thick that it wouldn't have leaked out before setting up anyway.

Figure 21-8. *Mold form coated with "liquid tape"*

3. Demold

In order to remove the silicone mold from the 3D-printed mold form, you will probably have to destroy the 3D-printed form completely. I tore mine completely apart. Some of the skulls had to be removed one by one. Overall it was pretty easy to demold, once I accepted that I was going to destroy the mold maker. Figure 21-9 and Figure 21-10 show the process.

Figure 21-9. *3D-printed mold form destruction!*

Figure 21-10. *Removing the skulls from the silicone mold*

Give the mold (Figure 21-11) a wash with soap and water in the sink, and I let it dry completely before attempting to cast chocolate. Water does bad things to chocolate when you are casting it.

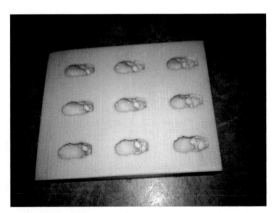

Figure 21-11. *Final silicone candy mold*

4. Add Slits to the Mold

Cut notches in the mold with a razor blade at the base of the skulls (see Figure 21-12). This will make it much easier to demold the chocolates.

Figure 21-12. *Slits added to the candy mold*

Figure 21-13. *Soy lecithin softgels*

I came across this by accident. After demolding many rounds of chocolates, the base of one of the skulls started to split a little from flexing the mold. I found that it had no impact on the quality of the chocolates coming out of the mold and it made it much easier to demold the chocolates. I cut little notches with a razor blade straight back perpendicular to the base of all of the skulls (the notches are only visible if I flex the mold) It makes it much easier to get them out of the mold.

5. Extract the Soy Lecithin from the Softgels

The soy lecithin will help with the demolding and has other benefits, such as acting as an emulsifier, when used in molding chocolate. I used Whole Foods soy lecithin softgels, which were available in the supplements isle.

I cut the softgels open and squeezed the oil out (Figure 21-13). I not very precisely used around two small handfuls of chocolate to the oil contained in four lecithin pills. Discard the softgels after squeezing out the oil.

6. Melt The Chocolate

I used white chocolate melts that do not require tempering. You could temper your own chocolate, but it is an extensive process in its own right and it is not covered here.

Heated chocolate melts in a double boiler. (If you have one, otherwise follow the alternate suggestions at the end of this section. I followed the detailed instructions from Chocoley (*http://chocoley.com/chocolate-candy-making-guide/melting-bada-bing-bada-boom-chocolate.htm*).

Keep in mind that temperature is very important. While the chocolate melts, use a thermometer to measure the temperature. Heat to between 100-105° F. Do not overheat the chocolate or you will ruin it for casting!

While melting the chocolate, add the small amount of lecithin that we obtained in the last step. I used two small handfuls of chocolate (I have small hands) to the oil obtained from four lecithin softgels. Gently stir the chocolate until it is consistently melted.

I also found that a tiny bit of cocoa powder can give the chocolate a more balanced taste. Just add a sprinkle to the batch.

I made a lot of chocolates, and I don't have a double boiler, so I experimented with several ways of melting the chocolate.

Bamboo Steamer + Glass Bowl

Fill a tall pot about halfway with water and place a bamboo steamer basket on top of the pot. Put a small glass mixing bowl on top of the bamboo steamer basket. Place your chocolate melts into the mixing bowl (Figure 21-14). I find it is best to do small batches. Add paper towels to the top of the steamer basket where it meets the glass bowl so that steam and chocolate do not mix.

Avoid getting water in the melted chocolate; it will dilute the mixture and it will not set properly when you try to cast it.

Figure 21-14. *Melting chocolate in a glass bowl and a bamboo steamer.*

Bamboo Steamer + Squeeze Bottle (Best Way)

After experimenting for an afternoon and melting several batches of chocolate, I found the following method to be the easiest to manage.

I kept the tall pot and the bamboo steamer set up, but switched from using a glass bowl to a chocolate squeeze bottle. I wrapped a dish towel around the bottle to block the rising steam (see Figure 21-15).

Figure 21-15. *Melting chocolate in the squeeze bottle.*

The squeeze bottle method made the process easier in two ways:

1. It was easy to fill the individual skulls in the mold and was much less messy than the other methods I tried.

2. After I filled the mold, I could put the squeeze bottle aside and it was easy to reheat later on top of the bamboo steamer. This eliminated messy and wasteful chocolate transfers using multiple containers and cut down on the dishes I needed to wash.

7. Let the Chocolate Cool

Before pouring the chocolate into the mold, you need to let the chocolate cool (Figure 21-16). The chocolate needs to cool down to about 96-98° F before you should start working with it. This will probably take longer than you think. Keep an eye on your thermometer and have your mold and tools ready. It pays to stand by the chocolate and wait until it cools; otherwise you may miss your window and it may cool too much to pour into the mold.

If you are going to pour the chocolate into a squeeze bottle or chocolate funnel, do it while it is cooling. Keep the thermometer in the chocolate so you can tell when it has cooled down enough to pour into the mold.

I tried a chocolate funnel before I settled on the squeeze bottle method.

carefully pour the chocolate from a small bowl into the holes in the mold.

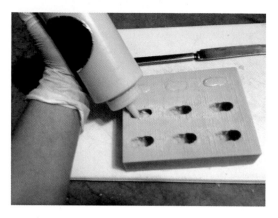

Figure 21-17. *Using a squeeze bottle to pour the chocolate into the mold*

After you pour the chocolate into the mold, use a spatula or butter knife to smooth out the top of the chocolate (Figure 21-18). Remove any excess on the top of the mold by scraping across the top.

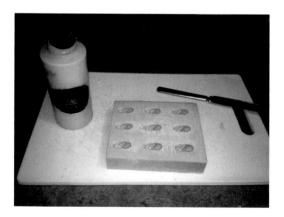

Figure 21-16. *Letting the chocolate cool—keep an eye on the thermometer!*

8. Pour the Chocolate into the Mold

Once the chocolate cools, you need to pour the chocolate into the mold. I recommend the squeeze bottle method (Figure 21-17), but you could also use a chocolate funnel or

Figure 21-18. *Smooth out the surface with a knife*

9. Put the Mold in the Refrigerator

Put the silicone mold containing the chocolate into the refrigerator until the chocolate solidifies (Figure 21-19). This will take about 20 minutes for small chocolates. It could take longer if you are casting a larger piece.

Figure 21-19. *Place in the fridge for 20 minutes*

10. While Waiting, Start Printing the Candy Trays

While you are waiting for the chocolate to set in the fridge, you can start printing your candy trays (Figure 21-20). If you don't have a 3D printer, you will need to plan for this or do with out the trays. Print the trays using PLA filament.

If you are concerned about whether PLA is food-safe for storage, put a small piece of wax or parchment paper beneath each skull.

Figure 21-20. *Printing PLA candy trays*

Here is the link to my candy tray files with four dividers: *http://www.thingiverse.com/thing:33432*. This parametric tray code was written by *acker*, who derived it from *hippie-gunnut*. I modified the parametric script to create a tray that would fit four of my chocolates perfectly.

11. Carefully Demold the Chocolate

The skull chocolates are a little difficult to get out of the mold. Wear candymaker's cotton gloves (Figure 21-21) to avoid marking the chocolate with your fingers.

Figure 21-21. *Demolding the chocolate with candy-maker's gloves*

Twist the rubber mold to loosen the chocolate. It will help to have small hands (I do and I am able to remove the chocolates without destroying them). I found that after loosening the chocolate by twisting the mold, it is easiest to twist out the face and rotate it out of the mold. Then grab the face and pull to remove the rest of the skull from the mold. You may smear the teeth a little, but overall it seems to work the best.

I have made many batches so far and I managed to get all of the chocolates out of the mold without destroying any of them (Figure 21-22).

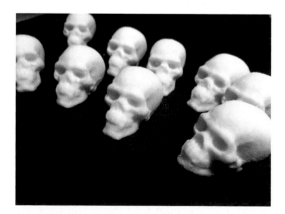

Figure 21-22. *Demolded chocolates*

Next, place the skulls into trays (Figure 21-23), put the trays in treat bags, and give them away to everyone!

Figure 21-23. *Place the chocolates in trays*

Anna Kaziunas France is the Digital Fabrication Editor of Maker Media. She's also the Dean of Students for the Global Fab Academy program and the co-author of Getting Started with MakerBot. *Formerly, she taught the "How to Make Almost Anything" rapid prototyping course in digital fabrication at the Providence Fab Academy. Learn more about her at her website (http://kaziunas.com) and check out her things at her Thingverse page (http://thingiverse.com/akaziuna).*

Printcrime | 22

A futuristic excerpt from *Overclocked: Stories of the Future Present.*

WRITTEN BY **CORY DOCTOROW**

The coppers smashed my father's printer when I was 8. I remember the hot, cling-film-in-a-microwave smell of it, and Da's look of ferocious concentration as he filled it with fresh goop, and the warm, fresh-baked feel of the objects that came out of it.

The coppers came through the door with truncheons swinging, one of them reciting the terms of the warrant through a bullhorn. One of Da's customers had shopped him. The ipolice paid in high-grade pharmaceuticals —performance enhancers, memory supplements, metabolic boosters. The kind of things that cost a fortune over the counter, the kind of things you could print at home, if you didn't mind the risk of having your kitchen filled with a sudden crush of big, beefy bodies, hard truncheons whistling through the air, smashing anyone and anything that got in the way.

They destroyed grandma's trunk, the one she'd brought from the old country. They smashed our little refrigerator and the purifier unit over the window. My tweetybird escaped death by hiding in a corner of his cage as a big, booted foot crushed most of it into a sad tangle of printer wire.

Da. What they did to him. When he was done, he looked like he'd been brawling with an entire rugby side. They brought him out the door and let the newsies get a good look at him as they tossed him in the car. All the while a spokesman told the world that my Da's organized-crime bootlegging opera-

195

tion had been responsible for at least 20 million in contraband, and that my Da, the desperate villain, had resisted arrest.

I saw it all from my phone, in the remains of the sitting room, watching it on the screen and wondering how, just how anyone could look at our little flat and our terrible, manky estate and mistake it for the home of an organized crime kingpin. They took the printer away, of course, and displayed it like a trophy for the newsies. Its little shrine in the kitchenette seemed horribly empty. When I roused myself and picked up the flat and rescued my poor peeping tweetybird, I put a blender there. It was made out of printed parts, so it would only last a month before I'd need to print new bearings and other moving parts. Back then, I could take apart and reassemble anything that could be printed.

By the time I turned 18, they were ready to let Da out of prison. I'd visited him three times—on my tenth birthday, on his fiftieth, and when Ma died. It had been two years since I'd last seen him and he was in bad shape. A prison fight had left him with a limp, and he looked over his shoulder so often it was like he had a tic. I was embarrassed when the minicab dropped us off in front of the estate, and tried to keep my distance from this ruined, limping skeleton as we went inside and up the stairs.

"Lanie," he said, as he sat me down. "You're a smart girl, I know that. You wouldn't know where your old Da could get a printer and some goop?"

I squeezed my hands into fists so tight my fingernails cut into my palms. I closed my eyes. "You've been in prison for 10 years, Da. Ten. Years. You're going to risk another 10 years to print out more blenders and pharma, more laptops and designer hats?"

He grinned. "I'm not stupid, Lanie. I've learned my lesson. There's no hat or laptop that's worth going to jail for. I'm not going to print none of that rubbish, never again." He had a cup of tea, and he drank it now like it was whisky, a sip and then a long, satisfied exhalation. He closed his eyes and leaned back in his chair.

"Come here, Lanie, let me whisper in your ear. Let me tell you the thing that I decided while I spent 10 years in lockup. Come here and listen to your stupid Da."

I felt a guilty pang about ticking him off. He was off his rocker, that much was clear. God knew what he went through in prison. "What, Da?" I said, leaning in close.

"Lanie, I'm going to print more printers. Lots more printers. One for everyone. That's worth going to jail for. That's worth anything."

Copy this story (*http://craphound.com/? p=573*).

Cory Doctorow (http://craphound.com) is a science fiction author, activist, journalist, blogger, co-editor of Boing Boing *(http://boingbo ing.net), and the author of the bestselling Tor Teen/HarperCollins UK novel* Little Brother. *His latest young adult novel is* Homeland, *and his latest novel for adults is* Rapture of the Nerds.

3D Printing Resources

WRITTEN BY **COLLEEN JORDAN, ERIC WEINHOFFER AND THE EDITORS OF MAKE**

Software for Makers

Creating things from atoms is better with bits.

Adafruit's 3D Printing Skill Badge

Declare to the world your proficiency in the world of 3D printing with this embroidered skill (*http://adafruit.com/products/490*) badge from Adafruit.

3D CAD

There are many useful and inexpensive software options for designing 3D models, for printing or otherwise.

123D Design
   http://123dapp.com

This is part of the free 123D suite of tools from Autodesk. You can model objects using its easy-to-learn interface, prepare your models for printing, export them to STL files, or send them directly to many popular fabrication companies. It includes a variety of popular 3D model creation, scanning, and sculpting apps, including the popular 123D Catch and 123D Design.

TinkerCAD
   http://tinkercad.com

Recently rescued by Autodesk, Tinkercad is a web-based modeling program. With a WebGL-enabled browser such as Google Chrome or Firefox, you can run Tinkercad's 3D user interface directly in your browser. Build up your design, save it online, and share it with others. You can also send files directly to popular 3D printing services or download STL files for printing yourself.

3DTin

http://www.3dtin.com

An in-browser tool that started out as a simple shape editor, with specific blocks that you can duplicate and manipulate to make models. Now it's become much more robust, with a multitude of modeling features.

OpenSCAD

http://openscad.org

If you like programming languages more than dragging and dropping, you might prefer OpenSCAD to the other modeling tools out there. Instead of drawing objects with your mouse, you program their shapes using lines of code. For example, cube([10,10,10]) will make a 10 mm cube appear onscreen. Using Boolean operators, you can combine, subtract, and intersect objects to create much more complex models using constructive solid geometry. Open-SCAD can export your scripts as STL models for 3D printing.

FreeCAD

http://free-cad.sourceforge.net

An open-source CAD program for Mac, Linux, and PC, built for product design and engineering. It's feature-rich and has a high learning curve. Check this out if you're interested in upgrading from a simpler program.

Sculptris

http://pixologic.com/sculptris

A free digital sculpting tool in which you create 3D models by interacting with "digital clay."

Cubify Invent

http://cubify.com/products/cubify_invent

An easy-to-learn 3D design tool from Cubify that has been optimized for 3D printing.

Cubify Sculpt

http://cubify.com/sculpt

Sculpt is an organic modeling tool from Cubify that enables digital sculpting and STL editing with mashup capability.

Trimble SketchUp

http://sketchup.com

Previously a Google project, SketchUp is now owned by Trimble. Out of the box, the free version isn't suitable for generating output for 3D printers, but there are many tutorials online for installing a plug-in that lets you export your designs to the required STL format.

Mesh Repair and Manipulation

MeshMixer

http://www.meshmixer.com

Purchased by Autodesk and recently integrated into their "experimental sandbox" of 123D apps, MeshMixer is a great tool for smoothing out, mashing up, repairing, or capping 3D scans and models.

netfabb

http://www.netfabb.com

Netfabb enables you to view and edit meshes and provides excellent repair and analysis capabilities for your STL files. It is currently available in both a free (basic) and paid versions.

Meshlab

http://meshlab.sourceforge.net

Meshlab can repair and edit meshes, but the learning curve can be a little daunting. Its poisson filter is great for smoothing surfaces when cleaning up 3D scans for printing. Its also a great mesh viewer and models views are easy to manipulate.

3D Printer Frontends

3D printer "frontends" are utilities to control your 3D printer. They all allow you to load STL files, arrange them on the build platform, slice the model, and send the files to the printer as G-code.

These programs all have an integrated "slicer" that chops your model up into layers to produce G-code, but can also send code generated from external slicers. See "Slicing Software" on page 200 for more information on individual slicers.

Repetier-Host
> http://repetier.com

> Unlike other frontends, Repetier-Host gives you three visualizations of your model: the 3D STL view, the layer-by-layer view of the G-code instructions that comprise the sliced model, and the real-time build view of each line of material as it's laid down. Slic3r is used by default, but Skeinforge is also available.

> Repetier-Host is available cross-platform, but the Windows and Linux versions have features not yet integrated into the Mac version (at the time of this writing).

Printrun/Pronterface
> http://reprap.org/wiki/printrun

> Printrun is made up of a suite of tools that use Pronterface as a frontend, without the data visualization features of Repetier-Host. You use Printrun to slice your model (by invoking Slic3r), then send your model to the printer. In addition to a graphical user interface,

Printrun includes command-line tools for working with print jobs.

MakerWare
> http://makerbot.com/makerware

MakerWare is the latest frontend printing software from MakerBot and is specifically designed for MakerBot 3D printers. It is easy to use, attractive, and intuitive. MakerWare is not open source, although you still have the option of using Skeinforge for slicing. MakerWare can load and slice more than one STL at a time, and there is an "auto layout" option for automatically arranging models on the build plate. The slicing profiles for both of the integrated slicers (the default MakerBot Slicer and Skeinforge, see "Slicing Software" on page 200) are fully editable.

Cura
> http://software.ultimaker.com

> The open source Cura printer control software was developed for use in the Ultimaker 3D printers, but can be set up to work with other machines. Cura is aesthetically beautiful and extremely easy to use for beginners, which is highly unusual in open source software. The latest versions of Cura have moved to a completely new slicing engine package, CuraEngine, (see "Slicing Software" on page 200) and are extremely fast.

ReplicatorG
> http://replicat.org

> ReplicatorG is an open source 3D printer frontend. It was originally created to support MakerBot printers, but they can also be used with other RepRap-based printers. ReplicatorG is getting less use these days, as many 3D printer vendors

now recommend Printrun or Repetier-Host.

However, a notable example of continued ReplicatorG use is in the Sailfish open source accelerated firmware project (*http://www.thingiverse.com/thing:32084*) by Jetty and Dan Newman for MakerBot printers.

Slicing Software

In order to print a 3D model, you first need to slice it using slicing software to generate the G-code to feed into your 3D printer. Slicers are sometimes integrated into the printer control software, but can also be standalone programs, like Slic3r and KISSlicer.

Slic3r
http://slic3r.org

Slic3r is a popular cross platform fast and open source slicer. It enables the user to store settings for both different printers and printing materials, making multi-machine prep easier.

KISSlicer
http://kisslicer.com

KISSlicer is a cross-platform G-code generator for desktop 3D printers that has been gaining popularity. It comes in free (single extruder) and "pro-grade" (multi-material, plating multiple ojects) versions.

CuraEngine
http://software.ultimaker.com

CuraEngine is an open source slicer developed specifically for Ultimaker, but it can also be used by other G-code-based printers. It is robust, powerful, and extremely fast—so fast it slices automatically each time you make a change to

your model or settings in Cura (see "3D Printer Frontends" on page 199).

MakerBot Slicer
http://www.makerbot.com/makerware

Formerly known as "Miracle Grue," the MakerBot Slicer is now the default MakerWare slicing option; it is fast, accurate, and provides nice print finishes. You can learn more about tuning your results with the MakerBot Slicer here: *http://www.makerbot.com/support/makerware/documentation/slicer*.

Skeinforge
http://reprap.org/wiki/Skeinforge

Skeinforge has been the slicing standard for years. It is written in Python and has many, cryptically named configurable settings. It can be slow and has begun to fall out of fashion among 3DP users. Skeinforge is still a possible slicing option for MakerWare and is still integrated into ReplicatorG and a few other custom printer control software applications.

MakerBot has created a handy "intro to Skeinforge" page: *http://www.makerbot.com/support/replicatorg/documentation/skeinforge*.

SFACT
http://reprap.org/wiki/Sfact

SFACT is a user friendly version of Skeinforge with lots of enhancements.

3D Model Repositories

Free 3D Model Downloads
Thingiverse
http://thingiverse.com

The Grand Bazaar of the 3D printing world, Thingiverse is where the DIY community shares 3D models, laser-cut files, and PCB layouts of their projects and printers.

Although under the ownership of MakerBot, it remains an important resource for hosting and distributing files for many of MakerBot's competitors. Users select from an array of Creative Commons licenses, including the option for public domain.

My Mini Factory

http://www.myminifactory.com

3D-printable models that have been vetted by the My Mini Factory team.

YouMagine

https://www.youmagine.com

Ultimaker's online community and 3D printable file sharing site.

CubeHero

https://cubehero.com

A free 3D model sharing site targeted at 3DP that now offers OpenSCAD previews. Some of the parts of the InMoov robot are located here.

Blender 3D Model Repository

http://blender-models.com

The Blender 3D Model Repository is another free site that hosts shared 3D model files. It's designed to be a resource for Blender 3D modeling software users to download, submit, and share their collective knowledge. Along with models, the site features user-submitted Blender tutorials.

Blend Swap

http://www.blendswap.com

Blend Swap is a proud community of 3D artists sharing their work and building the greatest Blender-based 3D assets

library, all free for personal and commercial use.

GrabCAD

http://grabcad.com

GrabCAD is the open engineering platform that gives you the tools, knowledge, and connections you need to build great products faster.

Trimble 3D Warehouse

http://www.sketchup.com/products/3D-warehouse

3D Content Central

http://www.3dcontentcentral.com

A Dassault Systèmes site that features 2D and 3D models of mechanical parts and a free service for locating, configuring, and downloading those parts.

McMaster-Carr

http://www.mcmaster.com/help/drawingsandmodels.asp

McMaster-Carr provide the 2D and 3D files for many of the products they sell.

Paid 3D Model Downloads

TurboSquid

http://turbosquid.com

TurboSquid is the leading paid 3D model repository, with over 200,000 models to download. Not optimized for 3D printing.

3D Burrito

http://3dburrito.com

3D Burrito features affordable downloads of 3D-printable models from designers like Bathsheba (Klein Bottle Opener) and Kid Mechanico of Modibot.

3DLT

http://3dlt.com

3DLT is a 3DP template marketplace that partners with independent designers to create the files offered on the website. You can either pay to download a file or access their partner network to have the file printed and shipped to you.

3docean
> *http://3docean.net*

3docean has lots of 3D models at reasonable prices, but they are not optimized for 3D printing.

3D Printer History

RepRap Family Tree
> *http://makezine.com/go/repraptree*

Shows the development of the RepRap 3D printer and its progeny.

Learn to Dial in Your Printer

If you get hands-on with a desktop 3D printer, there will be a point when you need general help, want to attempt something unusual or take your print quality to the next level. Depending on your printer model, there may be machine specific support forums provided by the manufacturer or you can investigate one of the following popular community sites.

RepRap.org
> *http://reprap.org*

A comprehensive source of information on RepRap 3D printers; including community resources, forums and details on numerous open source printer designs. MakerBot Operators Google Group::
> *https://groups.google.com/forum/#!fo rum/makerbot*

Your source for the latest on building and maintaining a deltabot style printer.

Jetty Firmware
> *http://www.thingiverse.com/thing:32084*

> *https://groups.google.com/forum/#!fo rum/jetty-firmware*

For discussion related to Jetty's and Dan Newman's Sailfish accelerated firmware for MakerBots. Sailfish firmware is a rapidly evolving open source derivative of Makerbot Industries firmware with additional onboard control options.

Deltabot Operators Google Group
> *https://groups.google.com/forum/#!fo rum/deltabot*

Your source for the latest on building and maintaining a deltabot style 3D printer.

Books

Design and Modeling for 3DP
Getting Started with MakerBot
> by Bre Pettis, Anna Kaziunas France, and Jay Shergill (MAKE)

Get a hands-on introduction to the world of personal fabrication with the popular MakerBot 3D printer. Not only will you learn how to operate, upgrade, and optimize your MakerBot, you'll also get guidelines on how to design, scan, and print your own prototypes. Read an excerpt on Chapter 6.

Design and Modeling for 3D Printing
> by Matthew Griffin (due out in early 2014; MAKE)

One to keep an eye out for, this book arms those entering the passionate, fast-moving field of 3D design with practical 3D design techniques and provides insights on designing for 3D printing through interviews and practical, hands-on examples.

3D CAD with Autodesk 123D: Designing for 3D Printing, Laser Cutting, and Personal Fabrication
Jesse Harrington Au (due out in early 2014, MAKE)

Learn to design for 3D printers and laser cutters with Autodesk's free and easy 123D tools. Autodesk's Maker Advocate shows you how to design objects from scratch, work with existing models, and scan real-world objects.

3D Printing and the Maker Movement

The Book on 3D Printing by Isaac Budmen and Anthony Rotolo (CreateSpace)

+ Artist/designer Isaac Budmen teamed up with digital tech expert/professor Anthony Rotolo to create a book that provides an approachable introduction to 3D printing. It covers machine technology, materials, and modeling software.

Practical 3D Printers—The Science and Art of 3D Printing
by Brian Evans (Apress)

Expert tinkerer Brian Evans examines a number of different 3D printers and the software and electronics that make them work. You'll learn how to calibrate, modify, and print objects with your 3D printer.

Makers: The New Industrial Revolution
by Chris Anderson (Random House)

Bestselling author and former *Wired* magazine editor Chris Anderson discusses open source design and desktop 3D printing, bringing you to the front lines of the new industrial revolution.

Fabricated: The New World of 3D Printing
by Hod Lipson and Melba Kurman (Wiley)

Fabricated provides readers with both practical and imaginative insights into the question "How will 3D printing technologies change my life?" based on research and interviews with experts from a broad range of industries.

3D Printing: The Next Industrial Revolution
by Christopher Barnatt (Crown Business)

An overview of 3D printing technologies and how they will transform our lives from professional futurist Christopher Barnatt.

Low-cost 3D Printing for Science, Education and Sustainable Development (Free Download)
edited by Enrique Canessa, Carlo Fonda, and Marco Zennaro

This book, available as a free download from *http://sdu.ictp.it/3D/book.html*, provides an overview of current research on 3D printing.

Whitepapers

Free whitepapers on copyright and 3D printing are available from the awesome team at Public Knowledge, whose mission is to preserve "the openness of the Internet and the public's access to knowledge, promoting creativity through balanced copyright."

What's the Deal with Copyright and 3D Printing?
by Michael Weinberg (Public Knowledge)

http://publicknowledge.org/ Copyright-3DPrinting

It Will Be Awesome if They Don't Screw It Up: 3D Printing, Intellectual Property, and the Fight over the Next Great Disruptive Technology

by Michael Weinberg (Public Knowledge)

http://publicknowledge.org/it-will-be-awesome-if-they-dont-screw-it-up

3DP News

The 3D Printing Industry
http://3dprintingindustry.com

A global resource for all things 3D printing.

3Ders.org
http://3Ders.org

3D printing news, updated daily.

MAKE
http://Makezine.com/3Dprinting

The latest 3D printing news, reviews, and projects.

Physical Destinations

Machine Access and Education

Makerspace Directory
http://makerspace.com/makerspace-directory

Makerspace.com is establishing a world-wide directory of makerspaces to encourage the growth of maker communities all over the globe.

Hackerspaces
http://hackerspaces.org/wiki/List_of_Hacker_Spaces

Many hackerspaces have 3D printers you can try out. Visit your local hacker-space to learn what equipment they have.

Fab Labs
http://www.fabfoundation.org/fab-labs
http://fab.cba.mit.edu/about/labs

Fab Labs are the educational outreach component of MIT's Center for Bits and Atoms (CBA), an extension of its research into digital fabrication and computation. They often function as independent makerspaces and share a common set of digital fabrication tools, including 3D printers.

TechShop
http://techshop.org

TechShop is a membership-based community workshop that has 3D printers and other tools available for members to use for a monthly or annual fee. There are TechShop locations in California, Michigan, Texas, Pennsylvania, and New York.

The 3D Printer Experience
http://www.the3dprinterexperience.com

The 3D Printer Experience offers classes on 3D modeling, scanning, and printing as well as 3D printing services.

316 North Clark St., Chicago, IL 60654

Brick and Mortar Stores

There are a few 3D printer stores sprinkled across the US and the UK. If you find yourself in one of these locales, don't pass up the opportunity to check these out!

Deezmaker
http://deezmaker.com

290 North Hill Ave. #5, Pasadena, CA 91106

GetPrinting3D Retail Store
  http://www.getprinting3d.com

  820 Davis St. Suite 111, Evanston, IL 60201

HoneyBee3D
  http://honeybee3d.com

  Montclair Village 6127 La Salle Avenue Oakland, CA 94611

iGo3D
  https://www.igo3d.com

  City Center Oldenburg, Germany

iMakr
  http://www.imakr.com

  79 Clerkenwell Road, Farringdon, London EC1R 5AR

MakerBot Retail Store
  http://makerbot.com/retail-store

  298 Mulberry St., New York, NY

Microcenter
  http://www.microcenter.com

  Various locations, 3D printing materials coming fall/winter 2013.

The 3D Printing Store
  http://the3dprintingstore.com

  4603 Monaco St., Denver, CO 80216

The Color Company
  http://www.color.co.uk

  27a Poland Street, London W1F 8QW

The UPS Store
  http://www.theupsstore.com/small-business-solutions/Pages/3D-printing.aspx

  As of July 31, 2013, the UPS Store began testing in-store 3D printing services at limited locations in San Diego, CA, Washington, DC, Frisco, TX, Menlo Park, CA, and Lisle, IL.

Printers, Filament, and Parts

http://makezine.com/go/suppliers
  A detailed directory of filament vendors that includes location, material specs, reviews, notes, and price.

Maker Shed
  http://makershed.com

  From our favorite 3D printers to filament and books, the Maker Shed has you covered.

Matter Geeks
  http://www.makergeeks.com

  Matter Geeks sells a huge variety of filament from PET, wood, stone, nylon, PVA, conductive, HIPS, flexible PLA and food-grade filaments in addition to the standard ABS and PLA. They also sell printers and provide both advice for printing and recommended temperatures for all their filaments.

Form Futura
  http://www.formfutura.com

  Form Futura sells flexible PLA, HIPS, nylon, wood, sandstone, ABS, and PLA filaments.

Filaco
  https://www.filaco.com

  HIPS, PLA and ABS filament.

Taulman 3D
  http://www.taulman3d.com

  Taulman 3D creates and sells several types of high strength nylon and PETT filament.

UltiMachine
http://ultimachine.com

RAMPS electronics, PLA, and ABS filament.

Faberdashery
http://faberdashery.co.uk

PLA filament in a variety of colors, by the meter.

Lulzbot
https://www.lulzbot.com

Suppliers of ABS, PLA, HIPS, Laywoo-D3 (wood), Taulman Nylon, and Polycarbonate.

DiamondAge
http://diamondage.co.nz

DiamondAge sells PLA in unique colors (like emerald and sapphire), as well as ABS filament.

Printbl
http://printbl.com

Authorized distributor of Diamond Age PLA. In addition they sell high impact (HIPLA) and impact-modified PLA (IMPLA).

Matter Hackers
https://www.matterhackers.com

ABS, PLA, wood, nylon and sandstone filament.

Amazon
http://www.amazon.com

Amazon.com now sells an array of 3D printing supplies, including filament, printers, and books on 3DP.

Inventables
https://www.inventables.com

Inventables, the "hardware store for designers," sells a large color variety of PLA; ABS, wood, and stone filaments; as well as 3D printers.

ProtoParadigm
http://protoparadigm.com

PLA and ABS in standard, unusual, and fluorescent colors.

QU-BD
http://qu-bd.com

Sells extruders, heated beds, and filament in a variety of colors in addition to 3D printers and CNC mills.

Deezmaker
http://deezmaker.com/store

Your source for syncromesh cables and pulleys and Budaschnozzle extruder nozzles, as well as Deezmaker printers.

MakerBot
http://www.makerbot.com

Sells MakerBot printers and MakerBot PLA and ABS filament.

Conferences

3D Print show (London, Paris, New York)
http://3dprintshow.com

Inside 3D Printing (New York, Chicago, San Jose, Singapore)
http://www.mediabistro.com/inside3dprinting/

Index

Symbols

We'd like to hear your suggestions for improving our indexes. Send email to index@oreilly.com.

Z

About the Compiler

Anna Kaziunas France is the Digital Fabrication Editor of Maker Media. She's also the Dean of Students for the Global Fab Academy program and the co-author of *Getting Started with MakerBot*. Formerly, she taught the "How to Make Almost Anything" rapid prototyping course in digital fabrication at the Providence Fab Academy. Learn more about her at her website (*http://kaziunas.com*) and check out her things at her Thingverse page (*http://thingiverse.com/akaziuna*).

Colophon

The cover and body font is BentonSans, the part title font is Camo Sans Heavy, the heading font is Serifa, and the code font is Bitstreams Vera Sans Mono.